SURFACE TENSION AND RELATED THERMODYNAMIC QUANTITIES OF AQUEOUS ELECTROLYTE SOLUTIONS

SURFACTANT SCIENCE SERIES

FOUNDING EDITOR

MARTIN J. SCHICK
1918–1998

SERIES EDITOR

ARTHUR T. HUBBARD
Santa Barbara Science Project
Santa Barbara, California

SURFACE TENSION AND RELATED THERMODYNAMIC QUANTITIES OF AQUEOUS ELECTROLYTE SOLUTIONS

Norihiro Matubayasi

Nagasaki University,
Nagasaki, Japan

CRC Press
Taylor & Francis Group
Boca Raton London New York

CRC Press is an imprint of the
Taylor & Francis Group, an **informa** business

CRC Press
Taylor & Francis Group
6000 Broken Sound Parkway NW, Suite 300
Boca Raton, FL 33487-2742

First issued in paperback 2019

ISBN-13: 978-1-4398-8087-6 (hbk)
ISBN-13: 978-0-367-37947-6 (pbk)

Library of Congress Cataloging-in-Publication Data

Matubayasi, Norihiro.
    Surface tension and related thermodynamic quantities of aqueous electrolyte solutions / Norihiro Matubayasi.
        pages cm -- (Surfactant science ; 157)
    Includes bibliographical references and index.
    ISBN 978-1-4398-8087-6 (hardcover : alk. paper)
    1. Surface tension. 2. Surface chemistry. 3. Thermodynamics. 4. Electrolyte solutions. I. Title.

QC183.M465 2014
530.4'275--dc23   2013027845

Visit the Taylor & Francis Web site at
http://www.taylorandfrancis.com

and the CRC Press Web site at
http://www.crcpress.com

Contents

Preface

This book is a comprehensive account of the properties of air/aqueous electrolyte surfaces, intended for graduate students and researchers in the field of colloid and interface science, with an emphasis on the contributions of simple ions to surface tension behavior, and the practical consequences. Surface tension provides a thermodynamic avenue for analyzing systems in equilibrium and formulating phenomenological explanations for the behavior of constituent molecules in the surface region. This approach generally lacks the ability to realize the underlying molecular mechanisms needed to understand them intuitively. To make up for this disadvantage, in general, thermodynamic studies involve systematic and extensive measurements. That is, thermodynamic studies are an experimental approach to obtaining vital fundamental knowledge known as "thermodynamic properties."

There are extensive experimental observations and established ideas regarding desorption of ions from the surfaces of aqueous salt solutions dating back to Langmuir's time. Langmuir was the first to calculate the thickness of the ion-free layer, and the variation of this thickness was evaluated by subsequent researchers. They recognized that the ion-free layer does not consist of pure water because some of the salt diffuses into it. Perhaps this theme has not attracted as much attention as it should because researchers have tended to focus on molecules having certain unique functions. However, interest in this theme has increased recently, with the development of new methods such as vibrational sum-frequency spectroscopy and molecular dynamics simulations.

Since the Wagner paper in 1924, it has been acknowledged that surface tension can be deduced from electrostatic forces, although the level of agreement with experimental surface tension measurements was inadequate. At the present time, a more successful discussion of the theory has emerged, which allows us to quantitatively calculate the distribution of ions in the surface region.

The main purpose of this book is to provide basic knowledge to conveniently unify the older and newer theories and measurements. Furthermore, there are many practical systems in the field of chemistry and related disciplines where one requires fundamental knowledge regarding ions in fluid/fluid systems. We hope that the surface tension and related thermodynamic quantities presented in this book will be helpful in understanding the basics of fluid/fluid interface in a wide variety of fields.

Extensive tables of surface tension data for aqueous electrolyte solutions useful in the calculation of thermodynamic quantities and in confirmation of theoretical consideration are given in appendices. I owe a great deal to my students and friends for these experimental data. I wish to thank them, and I express my gratitude for their support and assistance.

1 Introduction to Thermodynamic Consideration of Fluid/ Fluid Interface

When we take a look around our house or outside, we will see that all we can see is their shape. Everything is surrounded by other things, and there are boundaries between them. No matter how we remove the surface from bulky mass, we will always have a new surface. What we see with our eyes is the surface, but in reality it is mass observed by physical quantities. Then, what we study in a chemistry class are the properties of the bulk mass and little about the surfaces. Surface tension is also a physical quantity whose name is well known but neglected in general from the properties of the bulk mass. It was Gibbs who succeeded in developing thermodynamic relations, which involve surface tension. The main purpose of this first chapter is to provide a quick and easy guide to his theoretical insight and formulation of the boundary between fluids. It is intended for use as an overview for students who are not familiar with but have acquired basic physical chemistry.

1.1 PHASE AND PHASE BOUNDARIES

A single phase is a system whose intensive properties are independent of position. A system consisting of a single phase is called a homogeneous system. A system composed of two or more phases is called a heterogeneous system, and there are boundaries between adjacent phases. Some intensive properties such as temperature and chemical potential are kept constant through the boundaries, but we expect that there is a specific distribution in the densities of energy, mass, and others.

Although the atom's surface is visualized by an advanced technology such as scanning tunneling microscopy, our discussion of the effects of surface is not concerned with the dimension of an atom, a molecule, or an aggregate of a small number of molecules. The exact determination of the boundary is a matter of considerable difficulty, since it may be obscured and confused by the variety of boundaries. Actually, there is no criterion for using the common word "surface" in a phrase such as protein surface, and if one considers a single protein molecule in a solution from the standpoint of colloid science, one realizes the difficulties in assigning a precise meaning to it. According to Gibbs' phase rule, the number of degrees of freedom of a system or the variance of the system is defined by the number of phases and components. Then, when intensive properties of the system are described by the

1

combination of the intensive variables predicted by the phase rule, they consist of the properties of boundaries between adjacent phases. For example, when the aqueous solution of a protein is separated from pure water by a membrane that is permeable to water alone, the osmotic pressure π can be determined from the result of equating the chemical potential μ of pure water with that of the solution as

$$\mu_w^0(T, p) = \mu_w(T, p + \pi, x), \qquad (1.1)$$

where
  the subscript w refers to water as a component
  the superscript 0 refers to pure state

The osmotic pressure will not be observed for proteins large enough to satisfy the definition of phase, because the chemical potential of water in the dispersed system of such a large particle is a function of T and p only. However, it is impossible to make rigorous numerical distinction between the particles and the molecules upon their size. Some colloidal particles show dual properties. If the surface tension of the aqueous solution of surfactants is measured at narrow concentration range near their critical micelle concentration, it becomes practical to treat the micelles as the phase, because surface tension cannot be differentiable with respect to the concentration at the micelle formation. If the surface tension of the aqueous dispersions of the same micelles is measured in the wide concentration range, it seems practical to treat them as molecules, because surface tension decreases gradually with increasing concentration of micelles. It is apparent that the phase and the associated phase boundary should be defined by thermodynamics in macroscopic dimensions. The dimension of a physical surface of discontinuity between two phases should be much larger than that of a molecule when used in thermodynamic relations.

1.2   SURFACE OF DISCONTINUITY AND SURFACE REGION

The number of degrees of freedom of a two-phase two-component system with a phase boundary is two. Then the interfacial tension of a fluid/fluid boundary is fixed and has specific values for each fluid at fixed temperature and pressure. There is no doubt that interfacial tension is a macroscopic physical property of the boundary between fluid phases whose intensive properties are independent of position just up to the boundary, as long as thermodynamically defined interfacial tension is used. Gibbs used the term "surface of discontinuity" to denote this boundary, and he noted that the discontinuity does not imply the absolute one with mathematical precision. On the basis of the nature of the light reflected from the surface, it is supposed that transition between phases is confined to a layer of one or two molecules. Theoretical approaches have always been applied for a much wider continuous and gradual transition region designated as surface region. The surface is two-dimensionally homogeneous and infinite in extent but has a distinct discontinuity normal to the surface based on macroscopic observations.

1.3 PHENOMENA RESULTING FROM SURFACE TENSION

The shape of a liquid drop is easily deformed by an external force, but forms a spherical drop unless acted upon by the force. This change arises naturally from the lowering of the magnitude of the surface area, A, since all physical properties except the surface area are kept unchanged throughout the change in the shape. Thus the larger the surface tension γ of the liquid, the larger the potential of the surface, $\gamma \Delta A$. When this liquid droplet contacts with a solid surface, it either spreads on the surface of higher surface potential or rolls away on the surface of lower potential. At the beginning of the nineteenth century, Thomas Young described these phenomena by introducing concepts of surface tension, wettability, and contact angle (Young 1805). These three key words will well illustrate various phenomena caused by the boundary. If our face does not get wet, we could not wash our face. But even if it is wet, we could not wipe our face when the wettability of the towel is less than that of our face. If the legs of water striders get wet, they can not enjoy skating because surface tension pulls them into the ponds. Fish could not breathe without wetting their gills. The heterogeneous natural world is an ensemble of boundaries, and many phenomena in nature are relevant to wetting. The contact angle is a measure of the wetting, and it varies depending on the magnitude of the surface tension of boundaries between adjacent phases. Moreover, the surface tension varies depending on the temperature, pressure, and concentration of solute species dissolved in the bulk solutions.

Now let us consider how Young's equation illustrates the shape of a droplet on a flat solid surface. There are three boundaries: gas/liquid (G/L), liquid/solid (L/S), and gas/solid (G/S) surfaces as shown in Figure 1.1. At the point where these three types of interfaces meet, there is a balance of forces, which is named Young's equation, given as

$$\gamma_{G/S} = \gamma_{L/S} + \gamma_{G/L} \cos\theta. \tag{1.2}$$

The contact angle θ is an angle between $\gamma_{S/L}$ and tangent to $\gamma_{G/L}$ and is measured always in the liquid phase by convention. The wetting can be recognized by the contact angle approaching the zero value asymptotically. Finally, the drop will spread completely over the solid, and the line of contact will disappear as illustrated in the figure from left to right. This figure indicates that the G/S surface will be displaced by G/L and L/S surfaces when

$$\gamma_{L/S} + \gamma_{G/L} \leq \gamma_{G/S}. \tag{1.3}$$

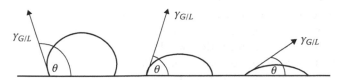

FIGURE 1.1 Schematic of drop shape and contact angle.

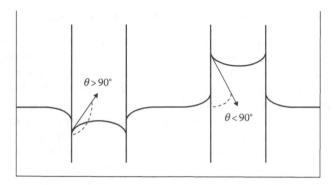

FIGURE 1.2 Wettability and shape of the meniscus.

The most familiar example of wettability shown in the textbook of colloid science is capillary action as demonstrated in Figure 1.2. When capillary tubes of hydrophilic glass and hydrophobic plastics are dipped into water, the meniscus in the glass tube is concave upward to the air phase while that in the hydrophobic tube is concave downward. Water spreads on the glass surface and pulls up the column of water. At the point where the three types of interfaces meet, the vertical component of the tangential of the surface tension of water is $\gamma\cos\theta$. The net upward force is $2\pi r\gamma\cos\theta$ and the force per unit area is $(2\gamma/r)\cos\theta$. The weight of the water column pulled by the meniscus is $\pi r^2 h\rho$ and the force per unit area is $h\rho g$, h and g being the height of the water column and the gravitational constant. The pressure difference between both sides of the meniscus is $h\rho g$. Let the pressures just above and below the meniscus be p^α and p^β, respectively. Then we have

$$p^\beta = p^\alpha - \frac{2\gamma}{r}\cos\theta. \tag{1.4}$$

If the radius of curvature is practically equal to the internal radius of the tube and the glass surface is completely wetted, Equation 1.4 reduces to

$$p^\alpha - p^\beta = \frac{2\gamma}{r}. \tag{1.5}$$

The pressure under the concave side of the meniscus is larger than the opposite side of the meniscus because of the force arising from surface tension. This equation is usually called Young–Laplace equation. The shape of the fluid is determined by the wettability and the surface tension, and their relation is derived in any textbook (Bikerman 1970). We know experimentally that the surface region consists of two or more molecular thicknesses, but the radius r in Equation 1.5 derived from physical considerations defines the position of the surface tension physically and mathematically.

  The shape of the oil lens formed on water surface is easily deformed by an external force, but forms a circular lens unless acted upon by the force. This process can be explained by the fact that the fluid flows out from the concave surface to the convex side because of the pressure difference. In this way, the Young–Laplace equation can be applied successfully to various phenomena. For example, when two water droplets with different diameters are connected by vapor, the water molecules in the smaller droplet will be moved to a larger one by means of the isopiestic distillation.

1.4 MECHANICAL AND THERMODYNAMIC DEFINITION OF SURFACE TENSION

The Young–Laplace equation we discussed earlier represents the mechanical balance between both sides of the curved boundary, which introduces the additional force of the surface tension. It is defined from a macroscopic viewpoint as the force across any element of the length of a line at the surface in a manner similar to that by which the pressure is defined. The first law of thermodynamics uses the concept of work in the discipline of mechanics and includes work other than volume changes such as electrical work.

1.4.1 WORK

Let us first consider the mechanical work done by the gravitational field, although the contribution is usually neglected in chemical thermodynamics. If a body of mass m falls in the gravitational field to the surface of the earth from the height Δh, the body will lose potential energy and do work upon the earth. We write the work W in the form

$$\Delta W = -mg\Delta h, \tag{1.6}$$

where g is the gravitational acceleration. Since the work is defined as the product of a force and a displacement, the force acted on the body with respect to the displacement of Δh is $-mg$. Similarly, for a displacement of volume, pressure is defined as the normal force per unit area exerted on a plane surface in the system, and we write

$$\Delta W = -p\Delta V. \tag{1.7}$$

If we add moles of pure substance into the system, the displacement will accompany the force that has the dimensions of joule per unit mol

$$\Delta W = \mu\Delta n \tag{1.8}$$

in which μ is the chemical potential.

Let us consider the work done when an area A of the surface between two adjacent phases changes by an amount ΔA. The force corresponding to the displacement ΔA is the surface tension; we can then write

$$\Delta W = \gamma\Delta A. \tag{1.9}$$

1.4.2 FUNDAMENTAL EQUATION

The first law of thermodynamics, an empirical deduction from the properties of work and heat, is the fundamental tool for understanding the energy of the system U. It shows that the increase in the energy of a system is the sum of the heat absorbed by the system ΔQ and the work done on the system ΔW. There are many verbal expressions of this rule in textbooks, but it can be most clearly expressed by the following relation:

$$\Delta U = \Delta W + \Delta Q. \tag{1.10}$$

It is well known that the second law of thermodynamics introduces the entropy for the reversible heat changes by the relation

$$\Delta Q = T\Delta S. \tag{1.11}$$

Substituting the work and heat changes into Equation 1.10, we obtain

$$dU = TdS - pdV + \gamma dA + \mu dn \tag{1.12}$$

with respect to the plane surface. In this equation, it is clear that U is considered to be a function of four variables including surface area and moles of component. Since the extensive properties are a linear homogeneous function of mass, we can write

$$U(\lambda S, \lambda V, \lambda A, \lambda n) = \lambda U(S, V, A, n), \tag{1.13}$$

where λ is a constant. Applying the Euler's theorem, the total derivative of this equation can be written in the form

$$dU = \left(\frac{\partial U}{\partial \lambda S}\right)d(\lambda S) + \left(\frac{\partial U}{\partial \lambda V}\right)d(\lambda V) + \left(\frac{\partial U}{\partial \lambda A}\right)d(\lambda A) + \left(\frac{\partial U}{\partial \lambda n}\right)d(\lambda n).$$

Division of this equation by $d\lambda$ results in

$$\frac{dU}{d\lambda} = \left(\frac{\partial U}{\partial \lambda S}\right)S + \left(\frac{\partial U}{\partial \lambda V}\right)V + \left(\frac{\partial U}{\partial \lambda A}\right)A + \left(\frac{\partial U}{\partial \lambda n}\right)n.$$

The derivative of the right side of Equation 1.13 with respect to λ is equal to this equation, and the equality holds for any values of λ. Then, for $\lambda = 1$,

$$U = TS - pV + \gamma A + \mu n. \tag{1.14}$$

Differentiating Equation 1.14 and combining with (1.12), we have

$$-SdT + Vdp - Ad\gamma - nd\mu = 0. \tag{1.15}$$

We define the Helmholtz free energy by

$$F = U - TS = -pV + \gamma A + \mu n. \tag{1.16}$$

The enthalpy and Gibbs free energy are

$$H = U + pV + \gamma A = TS + \mu n, \tag{1.17}$$

$$G = H - TS = \mu n. \tag{1.18}$$

The total differentials of these functions are

$$dF = d(U - TS) = -SdT - pdV + \gamma dA + \mu dn, \tag{1.19}$$

$$dH = d(U + pV - \gamma A) = TdS + Vdp - Ad\gamma + \mu dn, \tag{1.20}$$

$$dG = d(H - TS) = -SdT + Vdp - Ad\gamma + \mu dn. \tag{1.21}$$

Then, the surface tension is defined by

$$\gamma = \left(\frac{\partial U}{\partial A} \right)_{S,V,n} = \left(\frac{\partial F}{\partial A} \right)_{T,V,n}. \tag{1.22}$$

1.5 GIBBS' THERMODYNAMIC TREATMENT OF SURFACE

Gibbs' article titled "Influence of surfaces of discontinuity upon the equilibrium of heterogeneous masses. Theory of capillarity" may be the most cited article in surface science (Gibbs, 1875–1878). His thermodynamics is discussed in the textbooks of Kirkwood and Oppenheim (1961) with little modification and of Defay et al. (1966) with detailed analysis of the theory. However, a brief overview of Gibbs' thermodynamics is useful for the understanding of the thermodynamic relations of surface, because a lot of articles have referred to his adsorption equation without explanation.

1.5.1 Dividing Surface

Let us consider a closed system of, for example, phases α and β of volume V^α and V^β, respectively, which are separated by a spherical dividing surface (Ono and Kondo 1960). The location of this dividing surface is rather arbitrary and placed close to the surface of discontinuity. Figure 1.3 shows a conical vessel with solid angle of the cone ω in which phases α and β occupy the volume between radial coordinates r^α and R and between R and r^β, respectively. In this figure, the radial coordinate R shows the spherical mathematical surface between two fluids where mechanical equilibrium is established. Then, V^α and V^β are given respectively by

$$V^\alpha = \frac{\omega}{3} \left\{ R^3 - (r^\alpha)^3 \right\} \tag{1.23}$$

and

$$V^\beta = \frac{\omega}{3} \left\{ (r^\beta)^3 - R^3 \right\}. \tag{1.24}$$

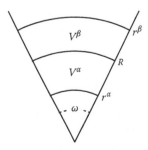

FIGURE 1.3 Dividing surface between two macroscopic phases α and β.

The area of the dividing surface is

$$A = \omega R^2. \tag{1.25}$$

Let us consider the change in the solid angle by $d\omega$ through reversible and isothermal displacements under constants r^α, r^β, and R; the work done by the system is given by

$$dW = \Omega d\omega = -p^\alpha dV^\alpha - p^\beta dV^\beta + \gamma dA. \tag{1.26}$$

By using (1.23), (1.24), and (1.26), we obtain

$$\gamma = \frac{R}{3}(p^\alpha - p^\beta) + \frac{K}{R^2}, \tag{1.27}$$

where

$$K = \Omega - \frac{p^\alpha}{3}(r^\alpha)^3 + \frac{p^\beta}{3}(r^\beta)^3. \tag{1.28}$$

Differentiating (1.27) with respect to R, Ono and Kondo obtained

$$(p^\alpha - p^\beta) = \frac{2\gamma}{R} + \left[\frac{d\gamma}{dR}\right]. \tag{1.29}$$

Since R is the dividing surface that separates mathematically fluid α from β under the condition that all the physical quantities of the system are fixed, $[d\gamma/dR]$ represents the change in the surface tension by the displacement of the arbitrary dividing surface. For a plane surface, this term is always zero, because R is infinite. It is evident that r in Equation 1.5 represents the surface so as to make

$$\left[\frac{d\gamma}{dR}\right] = 0. \tag{1.30}$$

This condition indicates that the surface tension does not depend on the location of the dividing surface that separates the system into two parts V^α and V^β. This particular dividing surface is called the surface of tension, where the Young–Laplace equation holds.

1.5.2   EXCESS OF EXTENSIVE THERMODYNAMIC QUANTITIES

Following Gibbs (1875–1878), let us consider the total energy of a heterogeneous system composed of two bulk phases α and β in equilibrium. These phases are homogeneous right up to the dividing surface considered earlier and have energies U^α and U^β, respectively. Gibbs defined excess energy U^σ, when on each side of the surface the density of energy had the same uniform value quite up to the surface. This definition of excess energy may be written as

$$U^\sigma = U - U^\alpha - U^\beta \tag{1.31}$$

under the condition that

$$0 = V - V^\alpha - V^\beta. \tag{1.32}$$

This condition (1.32) is a priori certain, as long as we consider the macroscopically defined physical surface of discontinuity. It is evident from molecular or microscopic consideration that the properties of molecules in contact with the dividing surface will be affected by the molecules of the adjacent phase even if the average values of those properties in the bulk phase are not affected from the macroscopic viewpoint. If there are no specific interactions between two adjacent masses, U^σ vanishes. Gibbs called U^σ and U^σ/A the superficial energy (surface energy) and superficial density of energy, respectively. In a similar manner, any extensive property Y of the system is described by the relation

$$Y^\sigma = Y - Y^\alpha - Y^\beta = Y - V^\alpha y^\alpha - V^\beta y^\beta \tag{1.33}$$

and its superficial density is defined as

$$y^\sigma = \frac{Y^\sigma}{A}. \tag{1.34}$$

Throughout this book, lowercase y represents the quantities per unit volume and y^σ represents the quantities per unit area.

1.5.3 Fundamental Equations for Plane Surfaces between Fluid Masses

Let us consider a two-phase two-component system with a plane interface. Following Gibbs' thermodynamic consideration of the surface, the variation of energy is given by

$$dU = TdS - pdV^\alpha - pdV^\beta + \gamma dA + \mu_1 dn_1 + \mu_2 dn_2. \tag{1.35}$$

Two similar equations for phases α and β are given respectively by

$$dU^\alpha = TdS^\alpha - pdV^\alpha + \mu_1 dn_1^\alpha + \mu_2 dn_2^\alpha \tag{1.36}$$

and

$$dU^\beta = TdS^\beta - pdV^\beta + \mu_1 dn_1^\beta + \mu_2 dn_2^\beta. \tag{1.37}$$

Making use of Euler's theorem, we can write

$$U = TS - pV^\alpha - pV^\beta + \gamma A + \mu_1 n_1 + \mu_2 n_2 \tag{1.38}$$

and for phases α and β,

$$U^\alpha = TS^\alpha - pV^\alpha + \mu_1 n_1^\alpha + \mu_2 n_2^\alpha \tag{1.39}$$

and

$$U^\beta = TS^\beta - pV^\beta + \mu_1 n_1^\beta + \mu_2 n_2^\beta. \tag{1.40}$$

In applying Equation 1.33, the superficial energy is found to be

$$dU^\sigma = TdS^\sigma + \gamma\, dA + \mu_1 dn_1^\sigma + \mu_2 dn_2^\sigma \tag{1.41}$$

and

$$U^\sigma = TS^\sigma + \gamma A + \mu_1 n_1^\sigma + \mu_2 n_2^\sigma. \tag{1.42}$$

Differentiation of Equation 1.42 and comparison with Equation 1.41 give the analog of the Gibbs–Duhem equation

$$S^\sigma dT + Ad\gamma + n_1^\sigma d\mu_1 + n_2^\sigma d\mu_2 = 0. \tag{1.43}$$

From this equation, we obtain

$$d\gamma = -s^\sigma dT - \Gamma_1 d\mu_1 - \Gamma_2 d\mu_2, \tag{1.44}$$

where Γ_i is a superficial density of component i defined as n_i^σ / A. Gibbs has presented a general equation for any number of components as

$$d\gamma = -s^\sigma dT - \Gamma_1 d\mu_1 - \Gamma_2 d\mu_2 - \cdots \tag{1.45}$$

called a fundamental equation of the surface of discontinuity (Gibbs, Eq. 508).

1.5.4   ONE-COMPONENT, TWO-PHASE SYSTEM

Let us consider the surface tension of the one-component system consisting of two homogeneous phases α and β in equilibrium. The Gibbs–Duhem equations for these phases are given respectively as

$$dp - c_1^\alpha d\mu_1 = s^\alpha dT \tag{1.46}$$

and

$$dp - c_1^\beta d\mu_1 = s^\beta dT. \tag{1.47}$$

In these equations, lowercase s and c represent the entropy and the moles of component per unit volume, respectively. The fundamental Equation 1.45 reduces to

$$d\gamma = -s^\sigma dT - \Gamma_1 d\mu_1. \tag{1.48}$$

Using Equations 1.46 and 1.47, we obtain (Gibbs, Eq. 578)

$$d\gamma = -\left[s^\sigma - \frac{\Gamma_1}{c_1^\alpha - c_1^\beta}(s^\alpha - s^\beta) \right] dT. \tag{1.49}$$

Gibbs pointed out that $\Gamma_1/(c_1^\alpha - c_1^\beta)$ represents the distance from the surface of tension to a dividing surface located so as to make the superficial density of the single component vanish. This suggestion can be easily explained using the schematic drawing of the surface region shown in Figure 1.4. It may be possible to consider that the real density distribution of component 1 gradually varies from the bulk of α to the bulk of β along the dotted line. However, we have supposed that both bulk phases are separated at the mathematical line that satisfies Equation 1.29. The superficial density of component 1 in reference to this dividing surface is given by

$$\Gamma_1 = \frac{n_1}{A} - l^\alpha c_1^\alpha - l^\beta c_1^\beta \tag{1.50}$$

where $l^\alpha = V^\alpha/A$ and $l^\beta = V^\beta/A$ are the lengths of phases α and β on the z coordinate. Similarly, we have

$$0 = \frac{n_1}{A} - (l^\alpha + \tau)c_1^\alpha - (l^\beta - \tau)c_1^\beta \tag{1.51}$$

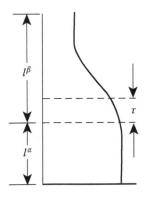

FIGURE 1.4 Schematic of the distance between the surface of tension and the dividing surface located so as to make the superficial density of the single component vanish.

with respect to the dividing surface located so as to make $\Gamma_1 = 0$. The τ in this equation, which is the distance between the two dividing surfaces, is obtained by combining the earlier two equations. As the heterogeneous system considered here consists of a pure liquid in equilibrium with its vapor, the location of the dividing surface is not probably very different from that located so as to make the superficial density of the component vanish. We can assume that $d\gamma/dT$ will give information about excess entropy per unit area.

1.5.5 Two-Component, Two-Phase System

First, we consider a system of components 1 and 2 that are practically immiscible in the liquid state and consisting of two pure liquid phases α and β in equilibrium.

The fundamental equation for this system is

$$d\gamma = -s^\sigma dT - \Gamma_1 d\mu_1 - \Gamma_2 d\mu_2. \tag{1.52}$$

Since the number of degrees of freedom of the system is two, it is desirable to eliminate one of the three variables. The Gibbs–Duhem equations for these phases are given respectively as

$$c_1^\alpha d\mu_1 + c_2^\alpha d\mu_2 = -s^\alpha dT + dp \tag{1.53}$$

and

$$c_1^\beta d\mu_1 + c_2^\beta d\mu_2 = -s^\beta dT + dp. \tag{1.54}$$

By the combination of these three equations under the condition that c_1^β and c_2^α are zero, the variation of surface tension between two immiscible fluids can be written (Gibbs, Eq. 580)

$$d\gamma = -\left(s^\sigma - \frac{\Gamma_1}{c_1^\alpha}s^\alpha - \frac{\Gamma_2}{c_2^\beta}s^\beta\right)dT - \left(\frac{\Gamma_1}{c_1^\alpha} + \frac{\Gamma_2}{c_2^\beta}\right)dp. \tag{1.55}$$

In a similar manner to that shown in the previous case, Γ_1/c_1^α and Γ_2/c_2^β are the distances from the surface of tension to two dividing surfaces of which one would make Γ_1 vanish, and the other Γ_2, respectively. Gibbs noted that $\left(\Gamma_1/c_1^\alpha + \Gamma_2/c_2^\beta\right)$ represents the distance between these two surfaces, or diminution of volume due to a unit of the surface of discontinuity.

Next, let us consider a two-component system consisting of a mixture of liquids that are in equilibrium with a gaseous mixture of them. In this case, it is generally convenient to place one more dividing surface so as to make the superficial density of component 1 vanish. This procedure immediately eliminates a variable as an independent variable in Equation 1.52, and the variation of the surface tension is (Gibbs, Eq. 514)

$$d\gamma = -s_{(1)}^\sigma dT - \Gamma_{2(1)} d\mu_2. \tag{1.56}$$

The subscript (1) is used by Gibbs for greater distinctness to denote the value of Γ_2 and s^σ as determined by the dividing surface placed so that $\Gamma_1 = 0$. It is evident that the validity of this equation rests upon the two dividing surfaces so as to make V^σ and n_1^σ vanish. The elimination of a variable can also be done without the use of the dividing surface but the use of the Gibbs–Duhem equations. We have two independent equations for two bulk phases α and β,

$$dp - c_1^\alpha \, d\mu_1 = s^\alpha \, dT + c_2^\alpha \, d\mu_2 \tag{1.57}$$

and

$$dp - c_1^\beta \, d\mu_1 = s^\beta \, dT + c_2^\beta \, d\mu_2. \tag{1.58}$$

From these equations, we obtain

$$d\mu_1 = -\frac{(s^\alpha - s^\beta)}{\left(c_1^\alpha - c_1^\beta\right)} \, dT - \frac{\left(c_2^\alpha - c_2^\beta\right)}{\left(c_1^\alpha - c_1^\beta\right)} \, d\mu_2. \tag{1.59}$$

Upon substitution of this in the fundamental Equation 1.52, we obtain $d\gamma$ as a function of T and μ_2 (Gibbs, Eq. 515)

$$d\gamma = -\left[s^\sigma - \frac{\Gamma_1}{c_1^\alpha - c_1^\beta}(s^\alpha - s^\beta) \right] dT - \left[\Gamma_2 - \frac{\Gamma_1}{c_1^\alpha - c_1^\beta}\left(c_2^\alpha - c_2^\beta\right) \right] d\mu_2. \tag{1.60}$$

By comparison with Equation 1.56, we find that

$$\Gamma_{2(1)} = \Gamma_2 - \frac{\Gamma_1}{c_1^\alpha - c_1^\beta}\left(c_2^\alpha - c_2^\beta\right) \tag{1.61}$$

and

$$s_{(1)}^\sigma = s^\sigma - \frac{\Gamma_1}{c_1^\alpha - c_1^\beta}(s^\alpha - s^\beta). \tag{1.62}$$

Gibbs pointed out that $\Gamma_1 / \left(c_1^\alpha - c_1^\beta\right)$ represents the distance between the surface of tension and the dividing surface that would make $\Gamma_1 = 0$. $\Gamma_{2(1)}$ is called the relative adsorption of component 2 with respect to component 1.

The relative adsorption has a special significance, because no definite location of the dividing surface would be required. Let us consider how this fact is presented under the condition that $V = V^\alpha + V^\beta$. The surface excess of the component i is defined by

$$n_1^\sigma = n_1 - V^\alpha c_1^\alpha - V^\beta c_1^\beta = n_1 - V c_1^\alpha + V^\beta \left(c_1^\alpha - c_1^\beta\right) \tag{1.63}$$

$$n_2^\sigma = n_2 - V^\alpha c_2^\alpha - V^\beta c_2^\beta = n_2 - V c_2^\alpha + V^\beta \left(c_2^\alpha - c_2^\beta\right). \tag{1.64}$$

Eliminating V^β between these equations, we obtain

$$\Gamma_2 - \frac{\Gamma_1}{c_1^\alpha - c_1^\beta}\left(c_2^\alpha - c_2^\beta\right) = \frac{1}{A}\left[\left(n_2 - Vc_2^\beta\right) - \left(n_1 - Vc_1^\alpha\right)\frac{c_2^\alpha - c_2^\beta}{c_1^\alpha - c_1^\beta}\right]. \tag{1.65}$$

This leads to the interesting result indicated earlier, since the right side of this equation consists of the quantities independent of the location of the dividing surface (Defay et al. 1966, Eq. 2.18). Of course, it is obvious that Γ_1 and Γ_2 are not found to be independent of the choice of the dividing surface. However, Equation 1.65 must hold for all combinations of Γ_1 and Γ_2 even when $\Gamma_1 = 0$. If we consider the superficial density of component 2 with reference to the dividing surface where superficial density of component 1 vanishes.

1.6 GIBBS' SURFACE MODEL

In the previous section, the approach to the thermodynamics of a heterogeneous system that includes surface contributions by Gibbs is reviewed briefly. It is found that his theoretical consideration is based on the deep consideration of the homogeneous phase and the phase rule. A macroscopic phase should be homogeneous just up to the surface of physical discontinuity, which is defined by the Young–Laplace equation. If we define the surface as a single phase, the number of degrees of freedom declines. For a pure surface, it is a common experience that variation of surface tension needs two independent variables. The characteristic feature of his approach is the introduction of two dividing surfaces. One divides the system into two homogeneous bulk phases by a macroscopic physical dividing surface so that the definition of the phase holds just below the surface, and the other is placed so as to make Γ_1 equal zero. Starting with a theoretically simple system such as the one-component two-phase system, he has developed the systematic method by which thermodynamic quantities can be quantitatively determined from the surface tension measurements as a function of temperature, pressure, and chemical potential of the component. It is found that the application to the determination of relative adsorption of solute using measurements of the surface tension of a solution in equilibrium with its vapor phase had been the most frequently used in the discipline of surface science. Application of Gibbs' thermodynamic development to experimental work is sometimes called Gibbs' model.

In order to apply Equation 1.56 to experimental work at the fluid/fluid interface, it is necessary to consider two points. One is to exchange the independent variable μ_2 for mole fraction of solute x_2^α. The total derivative of μ_2 will be given by

$$d\mu_2 = -\bar{s}_2 dT + \bar{v}_2 dp + \left(\frac{\partial \mu_2}{\partial x_2^\alpha}\right)_{T,p} dx_2^\alpha \tag{1.66}$$

where symbols $\bar{s}_2$ and $\bar{v}_2$ denote the partial molar entropy and volume of component 2, respectively. Since pressure is not used as an independent variable, we assume that Equation 1.66 can be written as

$$d\mu_2 = -\bar{s}_2 dT + \bar{v}_2\left(\frac{dp}{dT}\right)_{x_2} dT + \left(\frac{\partial \mu_2}{\partial x_2^\alpha}\right)_{T,p} dx_2^\alpha. \tag{1.67}$$

Substitution of Equation 1.67 in Equation 1.60 results in (Lewis and Randall 1961, Eq. 29.18):

$$dy = -\left[s_{(1)}^\sigma - \Gamma_{2(1)}\bar{s}_2 + \Gamma_{2(1)}\bar{v}_2\left(\frac{dp}{dT}\right)_{x_2} \right]dT - \Gamma_{2(1)}\left(\frac{\partial \mu_2}{\partial x_2^\alpha}\right)_T dx_2^\alpha. \tag{1.68}$$

If we assume that the solution is dilute and the vapor pressure is independent of the concentration, we obtain the well-known Gibbs' adsorption equation,

$$\Gamma_{2(1)} = -\frac{1}{RT}\left(\frac{\partial \gamma}{\partial \ln x_2^\alpha}\right)_T. \tag{1.69}$$

This relation is obtained by the application of the fundamental Equation 1.45 to a two-component system consisting of a solution that is in equilibrium with its vapor phase. When we try to apply this equation to the experimental measurements, it will be found necessary to introduce specially designed equipment. In general, surface tension measurements of solutions are made in the presence of an air phase instead of the vapor phase in equilibrium with solution under its vapor pressure. Experimental systems are usually the three-component (water, solute, air) two-phase (aqueous solution, air) systems.

Next let us consider the relative adsorption for the three-component system. The fundamental equation of this system in given by

$$dy = -s^\sigma dT - \Gamma_1 d\mu_1 - \Gamma_2 d\mu_2 - \Gamma_3 d\mu_3. \tag{1.70}$$

Gibbs–Duhem equations for phases α and β may be written respectively as

$$dp - c_1^\alpha d\mu_1 = s^\alpha dT + c_2^\alpha d\mu_2 + c_3^\alpha d\mu_3 \tag{1.71}$$

and

$$dp - c_1^\beta d\mu_1 = s^\beta dT + c_2^\beta d\mu_2 + c_3^\beta d\mu_3. \tag{1.72}$$

Using these Gibbs–Duhem equations, we have

$$d\mu_1 = -\frac{(s^\alpha - s^\beta)}{(c_1^\alpha - c_1^\beta)}dT - \frac{(c_2^\alpha - c_2^\beta)}{(c_1^\alpha - c_1^\beta)}d\mu_2 - \frac{(c_3^\alpha - c_3^\beta)}{(c_1^\alpha - c_1^\beta)}d\mu_3. \tag{1.73}$$

Substitution of this into Equation 1.70 results in

$$dy = -\left[s^\sigma - \frac{\Gamma_1(s^\alpha - s^\beta)}{c_1^\alpha - c_1^\beta} \right]dT - \left[\Gamma_2 - \frac{\Gamma_1(c_2^\alpha - c_2^\beta)}{c_1^\alpha - c_1^\beta} \right]d\mu_2 - \left[\Gamma_3 - \frac{\Gamma_1(c_3^\alpha - c_3^\beta)}{c_1^\alpha - c_1^\beta} \right]d\mu_3.$$

$$\tag{1.74}$$

Using Gibbs notation, we will write this as (Gibbs, Eq. 514):

$$dy = -s_{(1)}^{\sigma}dT - \Gamma_{2(1)}d\mu_2 - \Gamma_{3(1)}d\mu_3. \qquad (1.75)$$

Solutions are often very dilute, and since properties of solutes are different from that of solvents, it is questionable whether the chemical potential of the solvent of β phase would be treated as one of the independent variables. The relation with concentrations instead of chemical potentials as the independent variables may of course be deduced, but the formulae that express the surface tension will be less simple. This complexity may be the reason that Equation 1.69 developed for the two-component two-phase system has been favorably employed.

Usually, experiments are carried out using two immiscible solvents and a solute that is soluble in a phase, then Equation 1.69 can be used instead of (1.75). Further, the calculated $\Gamma_{2(1)}$ may have nearly the same value as the concentration in the surface region for the adsorbed surfactant film. However, subscript 2(1) has been used as a constant reminder that it is a surface excess quantity defined thermodynamically as an operationally useful quantity to describe the concentration in the surface region instead of the real number of moles of a solute molecule.

1.7 GUGGENHEIM'S SURFACE PHASE MODEL

Guggenheim (1940) has pointed out that, on the surface excess quantities introduced by Gibbs, thermodynamic quantities should be assigned to a mass with definite volume instead of a mathematical surface with no volume. He has defined the work done on the layer with definite thickness less than 10^{-8} m but with definite volume by V^{σ} by the sum of $-pdV^{\sigma}$ and γdA, and developed the equations for this layer (Guggenheim 1940, Guggenheim and Adam 1940, Guggenheim 1949). The quantity V^{σ} is no less important, but the definition of the work is the same in appearance as that of Gibbs, and the thermodynamic procedure and formulae developed are similar to those of Gibbs. From a theoretical point of view, the distinction between mathematical surface and molecular layers is not ambiguous, but it is uncertain if the volume of layers gives a satisfactory physical picture of the surface because the definition of the boundaries between layers and homogeneous phases is not clearly defined.

1.8 HANSEN'S CONVENTION

Following Gibbs' treatment, Hansen (1962) has developed the thermodynamic relations of the plane interface using a convention that eliminates chemical potentials of two components instead of a combination of pressure and chemical potential of a solvent. Let us consider a three-component two-phase system with a plane surface. From the Gibbs–Duhem equation of the entire system, we may write the differential of the surface tension in the form

$$Ady = -SdT + Vdp - n_1d\mu_1 - n_2d\mu_2 - n_3d\mu_3. \qquad (1.76)$$

Since there exist two more Gibbs–Duhem equations for two adjacent bulk phases α and β

$$-s^\alpha dT + dp - c_1^\alpha d\mu_1 - c_2^\alpha d\mu_2 - c_3^\alpha d\mu_3 = 0 \qquad (1.77)$$

and

$$-s^\beta dT + dp - c_1^\beta d\mu_1 - c_2^\beta d\mu_2 - c_3^\beta d\mu_3 = 0, \qquad (1.78)$$

two of the five variables T, p, μ_1, μ_2, μ_3 are not independent variables. Multiplying Equation 1.77 by V^α, (1.78) by V^β, and subtracting from Equation 1.76, we obtain

$$d\gamma = -s^\sigma dT + v^\sigma dp - \Gamma_1 d\mu_1 - \Gamma_2 d\mu_2 - \Gamma_3 d\mu_3, \qquad (1.79)$$

where s^σ, v^σ, and Γ_i are the superficial densities of entropy, volume, and the component i. From the experimental point of view, Hansen has eliminated mathematically μ_1 and μ_2 from the independent variables by setting Γ_1 and Γ_2 equal to zero. Then Equation 1.79 reduces to

$$d\gamma = -s^H dT + v^H dp - \Gamma_3^H d\mu_3. \qquad (1.80)$$

We used here superscript H instead of σ, which indicated that two dividing surfaces are defined so that the temperature and pressure can be used as the independent variables.

1.9 GIBBS' AND HANSEN'S CHOICE OF INDEPENDENT VARIABLES

In the preceding sections, the thermodynamic relations necessary for the interpretation of the derivative of surface tension have been reviewed. Let us now consider the distinction between Gibbs' and Hansen's choice of the independent variables. First we consider the two-component system with a plane surface where two immiscible liquid phases α and β coexist in equilibrium. For the two-component system, Equations 1.79 and 1.80 reduce to

$$d\gamma = -s^\sigma dT + v^\sigma dp - \Gamma_1 d\mu_1 - \Gamma_2 d\mu_2 \qquad (1.81)$$

and by setting Γ_1 and Γ_2 equal to zero, we have

$$d\gamma = -s^H dT + v^H dp. \qquad (1.82)$$

Thus, the surface tension increment occurring when the temperature is changed at fixed pressure is directly relevant to the superficial density of entropy characteristic of the two dividing surfaces.

Making use of Equations 1.53 and 1.54 and the condition that c_1^β and c_2^α are zero, Equation 1.81 reduces to

$$d\gamma = -\left(s^\sigma - \frac{\Gamma_1}{c_1^\alpha} s^\alpha - \frac{\Gamma_2}{c_2^\beta} s^\beta \right) dT + \left(v^\sigma - \frac{\Gamma_1}{c_1^\alpha} - \frac{\Gamma_2}{c_2^\beta} \right) dp. \qquad (1.83)$$

This algebraic manipulation is equivalent to that used in the derivation of Equation 1.55. Equations 1.55 and 1.83 are the explicit expressions for the surface tension of the two-component two-phase system when temperature and pressure are chosen as independent variables. Since Equations 1.82 and 1.83 are considered to have two dividing surfaces satisfying the condition that Γ_1 and Γ_2 equal zero, the coefficients of dT and dp are the superficial densities of the entropy and the volume, respectively. In contrast to Equation 1.83, Equation 1.55 is derived for a two-component system under the condition that $v^\sigma = 0$; the physical meaning of the coefficients of dT and dp in this equation is not simple.

Further, a comparison of Equations 1.75 and 1.80 suggests that the approximation used in Equation 1.67 is not needed when temperature and pressure are used as independent variables. The Gibbs adsorption equation for a two-component system has been used approximately for the measurements of the three-component system under atmospheric pressure and found to be fulfilled in all experimental work. It is assumed that the properties of a condensed phase are practically independent of pressure, and the specific effects of pressure are presumably negligible.

REFERENCES

Bikerman, J. J. 1970. *Physical Surfaces*. Academic Press, New York.

Defay, R., I. Prigogine, A. Bellmans, and D. H. Everett. 1966. *Surface Tension and Adsorption*. Longmans, London, U.K.

Gibbs, J. W. 1875–1878. Influences of surfaces of discontinuity upon the equilibrium of heterogeneous masses. Theory of capillarity. In *Collected Works*, Vol. 1, pp. 219–331, Ox Bow Press, Woodbridge, 1993.

Guggenheim, E. A. 1940. The thermodynamics of interfaces in systems of several components. *Trans. Faraday Soc.* 36: 397–412.

Guggenheim, E. A. 1949. *Thermodynamics*. North-Holland Publishing, Amsterdam, the Netherlands.

Guggenheim, E. A. and N. K. Adam. 1940. The thermodynamics of adsorption at the surface of solutions. *Proc. R. Soc. (Lond.)* A139: 218–236.

Hansen, R. S. J. 1962. Thermodynamics of interfaces between condensed phases. *J. Phys. Chem.* 66: 410–415.

Kirkwood, J. G. and I. Oppenheim. 1961. *Chemical Thermodynamics*. McGraw-Hill, New York.

Lewis, G. N. and M. Randall. 1961. *Thermodynamics*, 2nd edn., revised by K. S. Pitzer and. L. Brewer. McGraw-Hill, New York.

Ono, S. and S. Kondo. 1960. Molecular theory of surface tension in liquids. In *Handbuch der Physik*, ed. S. Flügge, Vol. X, pp. 134–278. Springer-Verlag, Berlin, Germany.

Young, T. 1805. An essay on the cohesion of fluids. *Philos. Trans. R. Soc. Lond.* 95: 65–87.

2 Basic Thermodynamic Relations for the Analysis of Fluid/Fluid Interface

In Chapter 1, we have reviewed the thermodynamic formalism, which accounts for the properties of a boundary between two fluid phases. However, the theoretical treatment of Gibbs (1875–1878) is described in reference to theoretically abstract systems instead of the practically useful problems. In this chapter, we consider the thermodynamic formulae that are conveniently applicable to experimental conditions commonly faced. The thermodynamic considerations presented here are limited to the plane interfaces based on a series of Motomura's seven articles published between 1978 and 1990 titled "Thermodynamic studies on adsorption at interfaces (Motomura 1978, 1980, Motomura et al. 1978b, 1978a, 1980, 1982, 1988, 1990)." The approach must start by an adequate definition for the dividing surfaces, which must then be followed by the choice of independent variables that satisfy the phase rule. This thermodynamic algorithm for obtaining the differentials of the surface tension is identical with that of Gibbs' shown in the previous chapter, but the formulae and interpretations of the thermodynamic quantities are useful for practical experimental work.

2.1 PHASE RULE

Since we are accustomed to the established thermodynamic equations that adhered to the Gibbs phase rule, we are less accustomed to the significance of the independent variables. A choice of the combination of independent variables is usually called as a convention, and it provides not only the experimental conditions but also the significance of the results. Let us first consider the Gibbs phase rule. By the first law of thermodynamics, it has been shown that the energy of a homogeneous system is expressed as a function of $S, V, n_1, \ldots, n_c$, and the differential of U is given by

$$dU = TdS - pdV + \sum_{i=1}^{c} \mu_i dn_i. \tag{2.1}$$

Using Euler's theorem, we obtain the Gibbs–Duhem equation

$$-SdT + Vdp - \sum_{i=1}^{c} n_i d\mu_i = 0 \tag{2.2}$$

that determines relationships among $c + 2$ intensive variables. Thus, we have a phase rule for a c-component one-phase system that $c + 1$ intensive variables are sufficient to describe an intensive property of the system. If this system is in equilibrium with another c-component homogeneous system, there exists one more Gibbs–Duhem equation. Using this new Gibbs–Duhem relation, we can eliminate one of the $c + 1$ variables. If p homogeneous phases are in equilibrium with each other, the number of variables needed to describe this heterogeneous system becomes

$$f = (c+1)-(p-1) = c+2-p. \tag{2.3}$$

It is obvious that the phase boundary has not been considered in the derivation of the Gibbs phase rule shown earlier.

Then, let us consider the c-component two-phase system with plane interface. The differential of U is given by

$$dU = TdS - pdV + \gamma\, dA + \sum_{i=1}^{c} \mu_i dn_i. \tag{2.4}$$

The Gibbs–Duhem equation of this system is given by

$$-SdT + Vdp - Ad\gamma - \sum_{i=1}^{c} n_i d\mu_i = 0. \tag{2.5}$$

The surface tension is described by $c + 2$ variables, but two Gibbs–Duhem equations of two adjacent phases can be used to reduce the number of variables. Then, we realize that the number of independent intensive variables that describe the system of plane surface becomes equal to that obtained by Equation 2.3.

2.2 FUNDAMENTAL EQUATION

Let us consider a two-phase system with plane interface consisting of three components a, b, and i.

The variation of energy is given by

$$dU = TdS - pdV + \gamma\, dA + \mu_a dn_a + \mu_b dn_b + \mu_i dn_i. \tag{2.6}$$

The Gibbs–Duhem equation of the system is

$$SdT - Vdp + Ad\gamma + n_a d\mu_a + n_b d\mu_b + n_i d\mu_i = 0. \tag{2.7}$$

Two Gibbs–Duhem equations for two bulk phases α and β in equilibrium are given respectively by

$$s^\alpha dT - dp + c_a^\alpha d\mu_a + c_b^\alpha d\mu_b + c_i^\alpha d\mu_i = 0 \tag{2.8}$$

and

$$s^\beta dT - dp + c_a^\beta d\mu_a + c_b^\beta d\mu_b + c_i^\beta d\mu_i = 0, \tag{2.9}$$

where s and c represent the entropy and the number of moles of a component per unit volume, respectively. When we consider the two-phase system, we use abbreviations such as phase α consists of solvent a and phase β consists of solvent b, respectively, throughout this book. Following Gibbs, we define the excess of the extensive properties Y^σ as

$$Y^\sigma = Y - Y^\alpha - Y^\beta. \tag{2.10}$$

Thus, for example, we write

$$S^\sigma = S - S^\alpha - S^\beta = S - V^\alpha s^\alpha - V s^\beta, \tag{2.11}$$

$$V^\sigma = V - V^\alpha - V^\beta, \tag{2.12}$$

and

$$n_j^\sigma = n_j - n_j^\alpha - n_j^\beta = n_j - V^\alpha c_j^\alpha - V^\beta c_j^\beta \quad j = a, b, i. \tag{2.13}$$

The surface densities of these quantities are

$$y^\sigma = \frac{Y^\sigma}{A} = \frac{Y}{A} - l^\alpha y^\alpha - l^\beta y^\beta. \tag{2.14}$$

Then,

$$s^\sigma = \frac{S^\sigma}{A} = \frac{S}{A} - l^\alpha s^\alpha - l^\beta s^\beta, \tag{2.15}$$

$$v^\sigma = \frac{V^\sigma}{A} = \frac{V}{A} - l^\alpha - l^\beta, \tag{2.16}$$

and

$$\Gamma_j = \frac{n_j^\sigma}{A} = \frac{n_j}{A} - l^\alpha c_j^\alpha - l^\beta c_j^\beta \quad j = a, b, i \tag{2.17}$$

where l^α and l^β are the lengths on the z coordinates of the two bulk phases with respect to two mathematical dividing surfaces, respectively. Multiplication of Equations 2.8 and 2.9 by l^α and l^β, respectively, and subtraction from Equation 2.7 yield

$$d\gamma = -s^\sigma dT + v^\sigma dp - \Gamma_a d\mu_a - \Gamma_b d\mu_b - \Gamma_i d\mu_i. \tag{2.18}$$

The number of degrees of freedom for a three-component two-phase system is three; two of the five variables on the right side of this fundamental equation depend on the other three variables. For the $c + 2$-component system, we have

$$d\gamma = -s^\sigma dT + v^\sigma dp - \Gamma_a d\mu_a - \Gamma_b d\mu_b - \sum_{i=1}^{c} \Gamma_i d\mu_i. \tag{2.19}$$

2.3   ONE-COMPONENT, TWO-PHASE SYSTEM

For a one-component two-phase system, Equation 2.18 reduces to

$$dy = -s^\sigma dT + v^\sigma dp - \Gamma_a d\mu_a. \tag{2.20}$$

In this section, we suppose that the system is composed of a liquid phase α of component a and its vapor phase β. It is clear by the phase rule that T, p, and μ_a cannot all be varied independently, since the number of degrees of freedom is one. There are three combinations of variables to be deleted.

2.3.1   SURFACE TENSION AS A FUNCTION OF TEMPERATURE

In the first place, let us eliminate p and μ_a from this relation using Gibbs–Duhem equations with respect to phases α and β, respectively:

$$c_a^\alpha d\mu_a - dp = -s^\alpha dT \tag{2.21}$$

and

$$c_a^\beta d\mu_a - dp = -s^\beta dT. \tag{2.22}$$

Thus, the surface tension of a pure liquid for temperature change becomes

$$dy = -\left(s^\sigma - \Gamma_a \frac{s^\alpha - s^\beta}{c_a^\alpha - c_a^\beta} - v^\sigma \frac{c_a^\alpha s^\beta - c_a^\beta s^\alpha}{c_a^\alpha - c_a^\beta} \right) dT. \tag{2.23}$$

An alternative expression for the quantity in the bracket can be derived using the definition of the surface excess of the volume and the concentration of component a. Using Equations 2.16 and 2.17, we readily obtain the expressions for l^α and l^β:

$$l^\alpha = \frac{1}{A} \frac{n_a - n_a^\sigma - c_a^\beta (V - V^\sigma)}{c_a^\alpha - c_a^\beta} \tag{2.24}$$

and

$$l^\beta = \frac{1}{A} \frac{c_a^\alpha (V - V^\sigma) - \left(n_a - n_a^\sigma \right)}{c_a^\alpha - c_a^\beta}. \tag{2.25}$$

Substituting these two equations in Equation 2.15, we obtain

$$s^\sigma - \Gamma_a \frac{s^\alpha - s^\beta}{c_a^\alpha - c_a^\beta} - v^\sigma \frac{c_a^\alpha s^\beta - c_a^\beta s^\alpha}{c_a^\alpha - c_a^\beta} = \frac{1}{A}\left(S - n_a \frac{s^\alpha - s^\beta}{c_a^\alpha - c_a^\beta} - V \frac{c_a^\alpha s^\beta - c_a^\beta s^\alpha}{c_a^\alpha - c_a^\beta} \right). \tag{2.26}$$

The left side of this equation must be independent of the position of the dividing surfaces because the right side contains quantities irrelevant to those surfaces. Then the rate of change of γ with respect to T given by Equation 2.23 is equal to the value of the right side of Equation 2.26 whatever the values of Γ_a and v^σ be. Let us choose two specific dividing surfaces so that Γ_a and v^σ are equal to zero, which is identical in form to that of the convention encountered in the treatment by Gibbs. It follows immediately by putting $\Gamma_a = 0$ and $v^\sigma = 0$ in both Equations 2.23 and 2.20 that the slopes of γ versus T plots represent the surface density of entropy with respect to the two dividing surfaces. Thus,

$$s^\sigma = \frac{d\gamma}{dT}. \tag{2.27}$$

2.3.2 Surface Tension as a Function of Pressure

In the second place, we eliminate variables T and μ_a from Equation 2.20 using two Gibbs–Duhem equations for gas and liquid phases in equilibrium. By this operation, the surface tension is expressed as a function of pressure in the form

$$d\gamma = \left[v^\sigma - \Gamma_a \frac{s^\alpha - s^\beta}{c_a^\beta s^\alpha - c_a^\alpha s^\beta} - s^\sigma \frac{c_a^\beta - c_a^\varepsilon}{c_a^\beta s^\alpha - c_a^\alpha s^\beta} \right] dp \tag{2.28}$$

and likewise, the terms in the square bracket are given by

$$v^\sigma - \Gamma_a \frac{s^\alpha - s^\beta}{c_a^\beta s^\alpha - c_a^\alpha s^\beta} - s^\sigma \frac{c_a^\beta - c_a^\alpha}{c_a^\beta s^\alpha - c_a^\alpha s^\beta} = \frac{1}{A} \left[V - n_a \frac{s^\beta - s^\alpha}{c_a^\alpha s^\beta - c_a^\beta s^\alpha} - S \frac{c_a^\alpha - c_a^\beta}{c_a^\alpha s^\beta - c_a^\beta s^\alpha} \right]. \tag{2.29}$$

This relation suggests that the numerical value in the bracket of Equation 2.28 is irrelevant to the position of the two dividing surfaces. If we place the dividing surfaces so that $\Gamma_a = 0$ and $s^\sigma = 0$, γ versus p plots can be used to determine the values of the surface density of volume v^σ with respect to the two dividing surfaces. Then when we choose pressure as an independent variable, surface tension measurements can be used to obtain

$$v^\sigma = \frac{d\gamma}{dp}. \tag{2.30}$$

2.3.3 Surface Tension as a Function of Chemical Potential

Finally, we can also eliminate T and p from Equation 2.20, and such a choice of the variables is usually impractical in an experimental work. We have

$$d\gamma = -\left(\Gamma_a - s^\sigma \frac{c_a^\beta - c_a^\alpha}{s^\beta - s^\alpha} + v^\sigma \frac{c_a^\beta s^\alpha - c_a^\alpha s^\beta}{s^\beta - s^\alpha} \right) d\mu_a. \tag{2.31}$$

By placing two dividing surfaces so that $s^\sigma = 0$ and $v^\sigma = 0$, we can evaluate Γ_a from the slope of the graph of γ versus μ_a plots.

2.4 QUASITHERMODYNAMIC RELATIONS

Before expanding the thermodynamic treatment to a two-component system, it is useful to consider the meaning of the surface excess quantities denoted by superscript σ. Quasithermodynamics is based on the assumption that the intensive thermodynamic quantities in the surface region are independent of the position and homogeneous in the direction laterally to the surface, but they are not in the direction perpendicular to the surface. In the surface region, it will be possible to define thermodynamic quantities at each microscopic point in the direction normal to the surface (Hill 1952, Ono and Kondo 1960). The Helmholtz free energy per unit volume at point z is given by

$$f(z) = -p_T(z) + c_a(z)\mu_a, \tag{2.32}$$

where

$p_T(z)$ is the tangential pressure at point z, which becomes the hydrostatic pressure
  p in the bulk phases
$c_a(z)$ is the local molar concentration of a (Ono and Kondo 1960, Motomura 1978)

The internal energy per unit volume is given by

$$u(z) = f(z) + Ts(z) \tag{2.33}$$

where $s(z)$ is the entropy per unit volume. Motomura (1978) has suggested that these local thermodynamic quantities are also expressed by the corresponding partial molar quantities in the following manner. For three components a, b, and i,

$$y(z) = c_a(z)\bar{y}_a(z) + c_b(z)\bar{y}_b(z) + c_i(z)\bar{y}_i(z), \tag{2.34}$$

where $\bar{y}_j(z)$ represents the local partial molar quantities of constituents. For a one-component system,

$$y(z) = c_a(z)\bar{y}_a(z). \tag{2.35}$$

The partial molar quantities of the one-component system are molar quantities, but we will use $\bar{y}_a$ to avoid confusing notation with quantities per unit volume y.

We shall now consider an expression for Γ_a defined by Equation 2.17. Let us consider a rectangular parallelepiped vessel in which the homogeneous gaseous and liquid phases of component a are separated across a surface region. We will focus on the surface region and divide this into three parts by two dividing planes normal to the z axis. Figure 2.1 illustrates an imaginary distribution of the molar concentration $c_a(z)$ along the z axis. Since surface excess quantities are defined with reference to the two dividing surfaces located at z^a and z^b, Γ_a will be given by

$$\Gamma_a = \int_{z^\alpha}^{z^\beta} c_a(z)dz - c_a^\alpha \int_{z^\alpha}^{z^a} dz - c_a^\beta \int_{z^b}^{z^\beta} dz. \tag{2.36}$$

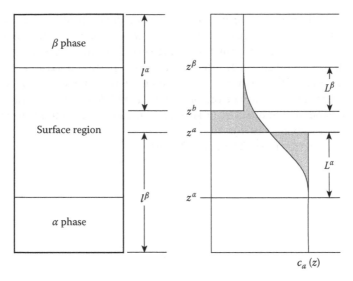

FIGURE 2.1   Schematic distribution of solvent a in the surface region with respect to two dividing surfaces.

The first term on the right side of the equation is found to represent the real number of moles of solvent a per unit area in the surface region which will be denoted by Γ_a^I. The second and third terms represent the area of a rectangle L^α by c_a^α and L^β by c_a^β, respectively. If we place the two dividing surfaces so that Γ_a equals zero, it is expressed as

$$\Gamma_a^I = \int_{z^\alpha}^{z^\beta} c_a(z)dz = c_a^\alpha \int_{z^\alpha}^{z^a} dz + c_a^\beta \int_{z^b}^{z^\beta} dz. \tag{2.37}$$

This choice of the dividing surface means that the shaded areas on the left and right sides of the solid line are equal. We next consider expressions for the surface density y^σ. We can write Equation 2.14 in the form

$$y^\sigma = \frac{V^I}{A} y - L^\alpha c_a^\alpha \bar{y}_a^\alpha - L^\beta c_a^\beta \bar{y}_a^\beta. \tag{2.38}$$

Here superscript I refers to the surface region. Using relation (2.35), we can express Equation 2.38 as

$$y^\sigma = \int_{z^\alpha}^{z^\beta} c_a(z)\bar{y}_a(z)dz - c_a^\alpha \bar{y}_a^\alpha \int_{z^\alpha}^{z^a} dz - c_a^\beta \bar{y}_a^\beta \int_{z^b}^{z^\beta} dz. \tag{2.39}$$

By comparing Equation 2.37 and Figure 2.1, we see that the first term on the right side of the earlier equation yields a quantity that is to be denoted as $\Gamma_a^I \langle \bar{y}_a^I \rangle$. Here $\langle \bar{y}_a^I \rangle$ refers to the mean partial molar quantity of the component a responsible for the surface region. The sum of the second and third terms gives the thermodynamic quantity y of bulk phases corresponding to the amount defined by the two dividing surfaces. Equation 2.39 suggests that y^σ represents a change in the partial molar quantity associated with the formation of the surface from the component a of the two bulk phases. In other words, this change in the partial molar quantities represents the variation of the quantities along with the adsorption of component a from both bulk phases into the surface region. Then we will denote this simply as Δy and call this thermodynamic quantity of surface formation or thermodynamic quantity of adsorption (Motomura 1978).

It is clear from Equations 2.23 and 2.28 that

$$\Delta s = -\frac{d\gamma}{dT} \tag{2.40}$$

and

$$\Delta v = \frac{d\gamma}{dp}. \tag{2.41}$$

2.5 SURFACE TENSION OF TWO-COMPONENT, TWO-PHASE SYSTEM

2.5.1 SURFACE TENSION AS A FUNCTION OF TEMPERATURE AND PRESSURE

Let us consider a system in which two partially miscible solvents a and b are in equilibrium and from the phase α and β respectively. Equation 2.18 reduces to

$$d\gamma = -s^\sigma dT + v^\sigma dp - \Gamma_a d\mu_a - \Gamma_b d\mu_b. \tag{2.42}$$

In order to express the surface tension in terms of two variables, the temperature and pressure will be the best choice of independent variables from an experimental convenience. To do this, we may use the following two Gibbs–Duhem equations for phases α and β:

$$c_a^\alpha d\mu_a + c_b^\alpha d\mu_b = -s^\alpha dT + dp \tag{2.43}$$

and

$$c_a^\beta d\mu_a + c_b^\beta d\mu_b = -s^\beta dT + dp. \tag{2.44}$$

Then, we can write the differential of the surface tension in the form

$$d\gamma = -\left(s^\sigma - \Gamma_a \frac{s^\alpha c_b^\beta - s^\beta c_b^\alpha}{c_a^\alpha c_b^\beta - c_b^\alpha c_a^\beta} - \Gamma_b \frac{s^\beta c_a^\alpha - s^\alpha c_a^\beta}{c_a^\alpha c_b^\beta - c_b^\alpha c_a^\beta} \right) dT$$

$$+ \left(v^\sigma - \Gamma_a \frac{c_b^\beta - c_b^\alpha}{c_a^\alpha c_b^\beta - c_b^\alpha c_a^\beta} - \Gamma_b \frac{c_a^\alpha - c_a^\beta}{c_a^\alpha c_b^\beta - c_b^\alpha c_a^\beta} \right) dp. \tag{2.45}$$

The coefficients of dT and dp are related to the quantities independent of the position of the dividing surface, respectively:

$$s^\sigma - \Gamma_a \frac{s^\alpha c_b^\beta - s^\beta c_b^\alpha}{c_a^\alpha c_b^\beta - c_b^\alpha c_a^\beta} - \Gamma_b \frac{s^\beta c_a^\alpha - s^\alpha c_a^\beta}{c_a^\alpha c_b^\beta - c_b^\alpha c_a^\beta} = \frac{1}{A}\left(S - n_a \frac{s^\alpha c_b^\beta - s^\beta c_b^\alpha}{c_a^\alpha c_b^\beta - c_b^\alpha c_a^\beta} - n_b \frac{s^\beta c_a^\alpha - s^\alpha c_a^\beta}{c_a^\alpha c_b^\beta - c_b^\alpha c_a^\beta} \right)$$

$$(2.46)$$

and

$$v^\sigma - \Gamma_a \frac{c_b^\beta - c_b^\alpha}{c_a^\alpha c_b^\beta - c_b^\alpha c_a^\beta} - \Gamma_b \frac{c_a^\alpha - c_a^\beta}{c_a^\alpha c_b^\beta - c_b^\alpha c_a^\beta} = \frac{1}{A}\left(V - n_a \frac{c_b^\beta - c_b^\alpha}{c_a^\alpha c_b^\beta - c_b^\alpha c_a^\beta} - n_b \frac{c_a^\alpha - c_a^\beta}{c_a^\alpha c_b^\beta - c_b^\alpha c_a^\beta} \right).$$

$$(2.47)$$

Then it will be convenient to consider that the temperature and pressure derivatives of the surface tension just give us the values of s^σ and v^σ by choosing the position of the two dividing surfaces so that Γ_a and Γ_b will be zero.

Let us consider the density of thermodynamic quantities based on the local formulation of local thermodynamics. Since Γ_a and Γ_b are defined to be zero, we have

$$\Gamma_a^I = \int_{z^\alpha}^{z^\beta} c_a(z)dz = c_a^\alpha \int_{z^\alpha}^{z^a} dz + c_a^\beta \int_{z^b}^{z^\beta} dz \qquad (2.48)$$

and

$$\Gamma_b^I = \int_{z^\alpha}^{z^\beta} c_b(z)dz = c_b^\alpha \int_{z^\alpha}^{z^a} dz + c_a^\beta \int_{z^b}^{z^\beta} dz. \qquad (2.49)$$

From Equation 2.34, we can write

$$y(z) = c_a(z)\bar{y}_a(z) + c_b(z)\bar{y}_b(z). \qquad (2.50)$$

Then

$$y^\sigma = \int_{z^\alpha}^{z^\beta} c_a(z)\bar{y}_a(z)dz - c_a^\alpha \bar{y}_a^\alpha \int_{z^\alpha}^{z^a} dz - c_a^\beta \bar{y}_a^\beta \int_{z^b}^{z^\beta} dz$$

$$+ \int_{z^\alpha}^{z^\beta} c_b(z)\bar{y}_b(z)dz - c_b^\alpha \bar{y}_b^\alpha \int_{z^\alpha}^{z^a} dz - c_b^\beta \bar{y}_b^\beta \int_{z^b}^{z^\beta} dz. \qquad (2.51)$$

By a similar argument to that of the one-component system, it seems better to call y^σ as "y of surface formation" and to denote it simply as Δy.

The Δy can be expressed in a simple form when solvents a and b are practically immiscible. Under the conditions, $c_a^\alpha \gg c_a^\beta$ and $c_b^\alpha \ll c_b^\beta$, Equations 2.48, 2.49, and 2.51 can be rewritten as

$$\Gamma_a^I = \int_{z^\alpha}^{z^\beta} c_a(z)dz = c_a^\alpha \int_{z^\alpha}^{z^a} dz, \tag{2.52}$$

$$\Gamma_b^I = \int_{z^\alpha}^{z^\beta} c_b(z)dz = c_b^\alpha \int_{z^\alpha}^{z^a} dz, \tag{2.53}$$

and

$$\Delta y = \int_{z^\alpha}^{z^\beta} c_a(z)\bar{y}_a(z)dz - c_a^\alpha \bar{y}_a^\alpha \int_{z^\alpha}^{z^a} dz + \int_{z^\alpha}^{z^\beta} c_b(z)\bar{y}_b(z)dz - c_b^\beta \bar{y}_b^\beta \int_{z^b}^{z^\beta} dz. \tag{2.54}$$

To understand the significance of Equation 2.54, we will rewrite this to indicate explicitly that Δy is the change in the thermodynamic quantity accompanied by the formation of an interface between phases α and β

$$\Delta y = \Gamma_a^I\left(\left\langle \bar{y}_a^I \right\rangle - \bar{y}_a^\alpha\right) + \Gamma_b^I\left(\left\langle \bar{y}_b^I \right\rangle - \bar{y}_b^\beta\right) \tag{2.55}$$

in which mean partial molar quantities of components a and b in the surface region $\left\langle \bar{y}_a^I \right\rangle$ and $\left\langle \bar{y}_b^I \right\rangle$ are defined respectively as

$$\Gamma_a^I\left\langle \bar{y}_a^I \right\rangle = \int_{z^\alpha}^{z^\beta} c_a(z)\bar{y}_a(z)dz \tag{2.56}$$

and

$$\Gamma_b^I\left\langle \bar{y}_b^I \right\rangle = \int_{z^\alpha}^{z^\beta} c_b(z)\bar{y}_b(z)dz. \tag{2.57}$$

Then, Equation 2.45 will take the simple form

$$d\gamma = -\Delta s dT + \Delta v dp. \tag{2.58}$$

2.5.2 Surface Tension as a Function of Temperature and Concentration

Let us next consider a new combination of independent variables, temperature and concentration of b in phase α. Equation 2.42 reduces to

$$d\gamma = -\left(s^\sigma - v^\sigma \frac{c_a^\alpha s^\beta - c_1^\beta s^\alpha}{c_a^\alpha - c_a^\beta} - \Gamma_a \frac{s^\alpha - s^\beta}{c_a^\alpha - c_a^\beta}\right)dT$$

$$-\left(\Gamma_b - v^\sigma \frac{c_a^\alpha c_b^\beta - c_a^\beta c_b^\alpha}{c_a^\alpha - c_a^\beta} - \Gamma_a \frac{c_b^\alpha - c_b^\beta}{c_a^\alpha - c_a^\beta}\right)d\mu_b \tag{2.59}$$

by using Gibbs–Duhem equations. Since the quantities in brackets are independent of the values of v^σ and Γ_a, we can write

$$d\gamma = -s^\sigma dT - \Gamma_b d\mu_b. \tag{2.60}$$

This equation is the same as Equation 1.56, although the selected dividing surfaces are not specified in the equation. In order to apply this to any experimental curve, it is convenient to express this in terms of the concentration x_b^α instead of the chemical potential μ_b. We can write μ_b in phase α in the form

$$d\mu_b = -\bar{s}_b^\alpha dT + \bar{v}_b^\alpha dp + \left(\frac{\partial \mu_b}{\partial x_b^\alpha}\right)_{T,p} dx_b^\alpha. \tag{2.61}$$

In order to substitute this into Equation 2.60, it might be needed to obtain the relation between p, T, and x_b^α when phase α is in equilibrium with β. This can be easily derived by the application of the description of heterogeneous systems in equilibrium (Kirkwood and Oppenheim 1961, Chapter 9). For phase β of this system, the Gibbs–Duhem equation can be written in the form

$$x_a^\beta \left(\bar{s}_a^\beta dT + \bar{v}_a^\beta dp - d\mu_a^\beta\right) + x_b^\beta \left(\bar{s}_b^\beta dT + \bar{v}_b^\beta dp - d\mu_b^\beta\right) = 0. \tag{2.62}$$

Substitution of Equation 2.61 and $d\mu_a$ with respect to T, p, and x_b^α into (2.62) results in the following equation:

$$dp = \frac{\left[x_a^\beta \left(\bar{s}_a^\beta - \bar{s}_a^\alpha\right) + x_b^\beta \left(\bar{s}_b^\beta - \bar{s}_b^\alpha\right)\right] dT + \left[x_a^\beta \left(\partial \mu_a / \partial x_b^\alpha\right) + x_b^\beta \left(\partial \mu_b / \partial x_b^\alpha\right)\right] dx_b^\alpha}{\left[x_a^\beta \left(\bar{v}_a^\beta - \bar{v}_a^\alpha\right) + x_b^\beta \left(\bar{v}_b^\beta - \bar{v}_b^\alpha\right)\right]}. $$

$$\tag{2.63}$$

Using Equations 2.61 and 2.63, we obtain

$$d\gamma = -\left(s^\sigma - \Gamma_b \bar{s}_b^\alpha + \Gamma_b \bar{v}_b^\alpha \frac{D_s}{D_v}\right) dT - \Gamma_b \left[\left(\frac{\partial \mu_b}{\partial x_b^\alpha}\right)_{T,p} + \bar{v}_b^\alpha \frac{D_b}{D_v}\right] dx_b^\alpha, \tag{2.64}$$

where

$$D_s = \left[x_a^\beta \left(\bar{s}_a^\beta - \bar{s}_a^\alpha\right) + x_b^\beta \left(\bar{s}_b^\beta - \bar{s}_b^\alpha\right)\right], \tag{2.65}$$

$$D_v = \left[x_a^\beta \left(\bar{v}_a^\beta - \bar{v}_a^\alpha\right) + x_b^\beta \left(\bar{v}_b^\beta - \bar{v}_b^\alpha\right)\right], \tag{2.66}$$

and

$$D_b = \left[x_a^\beta \left(\frac{\partial \mu_a}{\partial x_b^\alpha}\right) + x_b^\beta \left(\frac{\partial \mu_b}{\partial x_b^\alpha}\right)\right]. \tag{2.67}$$

If we consider the system in which phase α is an ideal dilute solution of solute b and phase β is a pure gaseous phase of a, Equation 2.64 takes the particularly simple form

$$d\gamma = -\left(s^\sigma - \Gamma_b \bar{s}_b^\alpha\right)dT - \Gamma_b\left(\frac{\partial \mu_b}{\partial x_b^\alpha}\right)_{T,p} dx_b^\alpha, \tag{2.68}$$

which is the same as Equation 1.68.

2.6 SURFACE TENSION OF THREE-COMPONENT, TWO-PHASE SYSTEM

Let us consider a dilute or moderately dilute solution of component i in solvent a in equilibrium with phase β composed of solvent b. It is clear by the application of the phase rule that three independent variables are selected out of five variables. One of the possible choices of the combination is temperature, pressure, and solute i. Following Gibbs, elimination of the variables can be done by assigning zero to the surface density of solvents, that is to say, by placing two dividing surfaces so that $\Gamma_a^\sigma = 0$ and $\Gamma_b^\sigma = 0$. Thus, Equation 2.18 reduces to

$$d\gamma = -s^H dT + v^H dp - \Gamma_i^H d\mu_i, \tag{2.69}$$

where superscript H is used to denote this particular choice. The quantities s^H, v^H, and Γ_i^H are respectively the surface densities of entropy, volume, and component provided that $\Gamma_a^\sigma = \Gamma_a^H = 0$ and $\Gamma_b^\sigma = \Gamma_b^H = 0$. For a $(2+c)$-component system, the fundamental equation is

$$d\gamma = -s^H dT + v^H dp - \sum_{i=1}^{c} \Gamma_i^H d\mu_i. \tag{2.70}$$

The elimination of variables can also be done by using Gibbs–Duhem equations of phases α and β. From Equations 2.8 and 2.9,

$$d\mu_a = -\frac{s^\alpha c_b^\beta - s^\beta c_b^\alpha}{c_a^\alpha c_b^\beta - c_b^\alpha c_a^\beta}dT + \frac{c_b^\beta - c_b^\alpha}{c_a^\alpha c_b^\beta - c_b^\alpha c_a^\beta}dp - \frac{c_i^\alpha c_b^\beta - c_i^\beta c_b^\alpha}{c_a^\alpha c_b^\beta - c_b^\alpha c_a^\beta}d\mu_i \tag{2.71}$$

and

$$d\mu_b = -\frac{s^\beta c_a^\alpha - s^\alpha c_a^\beta}{c_a^\alpha c_b^\beta - c_b^\alpha c_a^\beta}dT + \frac{c_a^\alpha - c_a^\beta}{c_a^\alpha c_b^\beta - c_b^\alpha c_a^\beta}dp - \frac{c_i^\beta c_a^\alpha - c_i^\alpha c_a^\beta}{c_a^\alpha c_b^\beta - c_b^\alpha c_a^\beta}d\mu_i. \tag{2.72}$$

By substituting these equations in Equation 2.18, we obtain

$$s^H = s^\sigma - \Gamma_a \frac{s^\alpha c_b^\beta - s^\beta c_b^\alpha}{c_a^\alpha c_b^\beta - c_b^\alpha c_a^\beta} - \Gamma_b \frac{s^\beta c_a^\alpha - s^\alpha c_a^\beta}{c_a^\alpha c_b^\beta - c_b^\alpha c_a^\beta},$$

(2.73)

$$v^H = v^\sigma - \Gamma_a \frac{c_b^\beta - c_b^\alpha}{c_a^\alpha c_b^\beta - c_b^\alpha c_a^\beta} - \Gamma_b \frac{c_a^\alpha - c_a^\beta}{c_a^\alpha c_b^\beta - c_b^\alpha c_a^\beta},$$

(2.74)

and

$$\Gamma_i^H = \Gamma_i - \Gamma_a \frac{c_i^\alpha c_b^\beta - c_i^\beta c_2^\alpha}{c_a^\alpha c_b^\beta - c_b^\alpha c_a^\beta} - \Gamma_b \frac{c_i^\beta c_a^\alpha - c_i^\alpha c_a^\beta}{c_a^\alpha c_b^\beta - c_b^\alpha c_a^\beta}.$$

(2.75)

The values of these surface densities are invariant with respect to the displacement of the dividing surface in a manner similar to Gibbs' relative adsorption. This fact can be shown from the consideration of the location of dividing surfaces. For solvents a and b, l^α and l^β are related to the surface densities by the relations

$$l^\alpha c_a^\alpha + l^\beta c_a^\beta = \frac{n_a}{A} - \frac{n_a^\sigma}{A}$$

(2.76)

and

$$l^\alpha c_b^\alpha + l^\beta c_b^\beta = \frac{n_b}{A} - \frac{n_b^\sigma}{A}.$$

(2.77)

Solving these equations for l^α and l^β, we have

$$l^\alpha = \frac{\left(n_a - n_a^\sigma\right)c_b^\beta - \left(n_b - n_b^\sigma\right)c_a^\beta}{A\left(c_a^\alpha c_b^\beta - c_a^\beta c_b^\alpha\right)}$$

(2.78)

and

$$l^\beta = \frac{\left(n_b - n_b^\sigma\right)c_a^\alpha - \left(n_a - n_a^\sigma\right)c_b^\alpha}{A\left(c_a^\alpha c_b^\beta - c_a^\beta c_b^\alpha\right)}.$$

(2.79)

Substituting these l^α and l^β into Equations 2.15 through 2.17, we obtain

$$s^\sigma - \frac{s^\alpha c_b^\beta - s^\beta c_b^\alpha}{\left(c_a^\alpha c_b^\beta - c_a^\beta c_b^\alpha\right)}\Gamma_a - \frac{s^\beta c_a^\alpha - s^\alpha c_a^\beta}{\left(c_a^\alpha c_b^\beta - c_a^\beta c_b^\alpha\right)}\Gamma_b$$

$$= \frac{1}{A}\left[S - \frac{s^\alpha c_b^\beta - s^\beta c_b^\alpha}{\left(c_a^\alpha c_b^\beta - c_a^\beta c_b^\alpha\right)}n_a - \frac{s^\beta c_a^\alpha - s^\alpha c_a^\beta}{\left(c_a^\alpha c_b^\beta - c_a^\beta c_b^\alpha\right)}n_b\right]$$

(2.80)

$$v^\sigma - \frac{\left(c_b^\beta - c_b^\alpha\right)}{\left(c_a^\alpha c_b^\beta - c_a^\beta c_b^\alpha\right)}\Gamma_a - \frac{\left(c_a^\alpha - c_a^\beta\right)}{\left(c_a^\alpha c_b^\beta - c_a^\beta c_b^\alpha\right)}\Gamma_b$$

$$= \frac{1}{A}\left[V - \frac{\left(c_b^\beta - c_b^\alpha\right)}{\left(c_a^\alpha c_b^\beta - c_a^\beta c_b^\alpha\right)}n_a - \frac{\left(c_a^\alpha - c_a^\beta\right)}{\left(c_a^\alpha c_b^\beta - c_a^\beta c_b^\alpha\right)}n_b\right] \qquad (2.81)$$

and

$$\Gamma_i - \frac{c_i^\alpha c_b^\beta - c_i^\beta c_b^\alpha}{\left(c_a^\alpha c_b^\beta - c_a^\beta c_b^\alpha\right)}\Gamma_a - \frac{c_i^\beta c_a^\alpha - c_i^\alpha c_a^\beta}{\left(c_a^\alpha c_b^\beta - c_a^\beta c_b^\alpha\right)}\Gamma_b$$

$$= \frac{1}{A}\left[n_i - \frac{c_i^\alpha c_b^\beta - c_i^\beta c_b^\alpha}{\left(c_a^\alpha c_b^\beta - c_a^\beta c_b^\alpha\right)}n_a - \frac{c_i^\beta c_a^\alpha - c_i^\alpha c_a^\beta}{\left(c_a^\alpha c_b^\beta - c_a^\beta c_b^\alpha\right)}n_b\right]. \qquad (2.82)$$

The left sides of these equations are respectively the coefficients of Equation 2.69 s^H, v^H, and Γ_i^H, and each term contains a quantity inherent in the dividing surface; however, on the right side of these equations, the quantities in square brackets are not related to inherent properties of the surface. It is to be noted that s^H, v^H, and Γ_i^H are the surface excess quantities with respect to the two dividing surfaces so that $\Gamma_a = 0$ and $\Gamma_b = 0$, respectively.

2.7 SURFACE TENSION AS A FUNCTION OF TEMPERATURE, PRESSURE, AND CONCENTRATION

Equation 2.69 describes the surface tension as a function of T, p, and μ_i. In applying this, it is probably best to use mole fraction instead of the chemical potential of the solute. The total differential of the chemical potential of component i in phase α is given by

$$d\mu_i^\alpha = -\bar{s}_i^\alpha dT + \bar{v}_i^\alpha dp + \left(\frac{\partial \mu_i}{\partial x_a^\alpha}\right)dx_a^\alpha + \left(\frac{\partial \mu_i}{\partial x_i^\alpha}\right)dx_i^\alpha, \qquad (2.83)$$

where a bar over the thermodynamic quantity is used to represent the partial molar quantity. When phase α is in equilibrium with phase β, one degree of freedom decreases, and variables T, p, and x_i^α will be the best combination to describe $d\mu_i^\alpha$. We can use the following three equilibrium conditions in order to derive dx_a^α as a function of T, p, and x_i^α (Kirkwood and Oppenheim 1961, Chapter 9). At equilibrium,

$$d\mu_a^\alpha\left(T, p, x_a^\alpha, x_i^\alpha\right) = d\mu_a^\beta\left(T, p, x_a^\beta, x_i^\beta\right), \qquad (2.84)$$

$$d\mu_b^\alpha\left(T, p, x_a^\alpha, x_i^\alpha\right) = d\mu_b^\beta\left(T, p, x_a^\beta, x_i^\beta\right), \qquad (2.85)$$

and

$$d\mu_i^\alpha\left(T, p, x_a^\alpha, x_i^\alpha\right) = d\mu_i^\beta\left(T, p, x_a^\beta, x_i^\beta\right). \qquad (2.86)$$

Thus the change in x_a^α can be written as

$$dx_a^\alpha = -\frac{D_s}{D_a}dT + \frac{D_v}{D_a}dp - \frac{D_i}{D_a}dx_i^\alpha, \qquad (2.87)$$

where

$$D_j = x_a^\beta \left(\frac{\partial \mu_a^\alpha}{\partial x_j^\alpha}\right) + x_b^\beta \left(\frac{\partial \mu_b^\alpha}{\partial x_j^\alpha}\right) + x_i^\beta \left(\frac{\partial \mu_i^\alpha}{\partial x_j^\alpha}\right) \qquad (2.88)$$

and

$$D_y = x_a^\beta \left(\overline{y}_a^\beta - \overline{y}_a^\alpha\right) + x_b^\beta \left(\overline{y}_b^\beta - \overline{y}_b^\alpha\right) + x_i^\beta \left(\overline{y}_i^\beta - \overline{y}_i^\alpha\right), \quad y = s, v. \qquad (2.89)$$

Substitution of Equation 2.87 into (2.83) yields

$$d\mu_i^\alpha = -\left[\overline{s}_i^\alpha + \frac{D_s}{D_a}\left(\frac{\partial \mu_i}{\partial x_a^\alpha}\right)\right]dT + \left[\overline{v}_i^\alpha + \frac{D_v}{D_a}\left(\frac{\partial \mu_i}{\partial x_a^\alpha}\right)\right]dp + \left[\left(\frac{\partial \mu_i}{\partial x_i^\alpha}\right) - \frac{D_i}{D_a}\left(\frac{\partial \mu_i}{\partial x_a^\alpha}\right)\right]dx_a^\alpha.$$
$$\qquad (2.90)$$

When the solutions are dilute so that the changes in the activity coefficients are negligibly small, the derivative of the chemical potential with respect to x_a^α can be considered to be zero. Then, Equation 2.90 reduces to

$$d\mu_i^\alpha = -\overline{s}_i^\alpha dT + \overline{v}_i^\alpha dp + \left(\frac{\partial \mu_i}{\partial x_i^\alpha}\right)_{T,p} dx_i^\alpha. \qquad (2.91)$$

Further, we can use Equation 2.91 for the system where solvents of phases α and β are practically immiscible with each other and solutes are insoluble in the solvent b. For such a system, Equation 2.69 as we may write

$$d\gamma = -\Delta s dT + \Delta v dp - \Gamma_i^H \left(\frac{\partial \mu_i}{\partial x_i^\alpha}\right)_{T,p} dx_i^\alpha, \qquad (2.92)$$

where Δs and Δv are defined respectively as

$$\Delta s = s^H - \Gamma_i^H \overline{s}_i^\alpha \qquad (2.93)$$

and

$$\Delta v = v^H - \Gamma_i^H \overline{v}_i^\alpha. \qquad (2.94)$$

We may call Δs and Δv the entropy and volume of adsorption of solute i from phase α, respectively (Motomura 1978). The magnitude of these values can be evaluated experimentally by

$$\Delta s = -\left(\frac{\partial \gamma}{\partial T}\right)_{p,x_i^\alpha} \tag{2.95}$$

and

$$\Delta v = \left(\frac{\partial \gamma}{\partial p}\right)_{T,x_i^\alpha}. \tag{2.96}$$

If the solution is ideal,

$$\mu_i = \mu_i^0(T,p) + RT \ln x_i^\alpha, \tag{2.97}$$

and consequently,

$$\Gamma_i^H = -\frac{x_i^\alpha}{RT}\left(\frac{\partial \gamma}{\partial x_i^\alpha}\right)_{T,p}. \tag{2.98}$$

This equation is of the same form as the well-known Gibbs adsorption equation derived for the two-component system, because we assumed that phases α and β are practically immiscible.

2.8 QUASITHERMODYNAMIC RELATIONS FOR ADSORPTION

Let $c_j(z)$ be the concentration of component j in the number of moles per unit volume at a given point z. We have drawn typical imaginary curves for two solvents a and b and a solute i, respectively, in Figure 2.2, where two homogeneous phases are separated by a surface region. The regions below the plane z^α and above the plane z^β are the homogeneous phases α and β, respectively. Planes z^a and z^b show the two dividing surfaces where $\Gamma_a = 0$ and $\Gamma_b = 0$, respectively. In the surface region, $c_j(z)$ varies continuously from phase α to β along the thick solid line, and the region on the left side of the line between the planes z^α and z^β shows the number of moles of the components responsible for the surface region.

Referring to this figure, we now consider the concentrations of solvents a and b in the surface region. The surface densities of the solvents defined by Equation 2.17 can be written in the form

$$\Gamma_a = \int_{z^\alpha}^{z^\beta} c_a(z)dz - c_a^\alpha \int_{z^\alpha}^{z^a} dz - c_a^\beta \int_{z^b}^{z^\beta} dz \tag{2.99}$$

and

$$\Gamma_b = \int_{z^\alpha}^{z^\beta} c_b(z)dz - c_b^\alpha \int_{z^\alpha}^{z^a} dz - c_b^\beta \int_{z^b}^{z^\beta} dz. \tag{2.100}$$

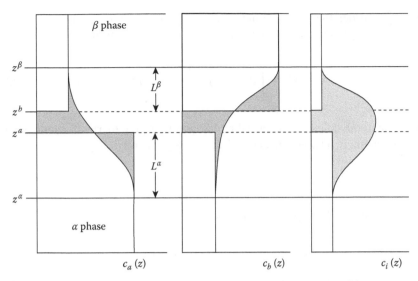

FIGURE 2.2 Schematic distributions of solvents a and b and solute i with respect to two dividing surfaces.

These are not real densities of the solvents in the surface region, since Gibbs defined these surface densities as the excess quantity. The two dividing surfaces are defined so as to make the surface excess of solvents vanish. Thus the real densities of the solvents are

$$\Gamma_a^I = \int_{z^\alpha}^{z^\beta} c_a(z)dz = c_a^\alpha \int_{z^\alpha}^{z^a} dz + c_a^\beta \int_{z^b}^{z^\beta} dz = L^\alpha c_a^\alpha + L^\beta c_a^\beta \tag{2.101}$$

and

$$\Gamma_b^I = \int_{z^\alpha}^{z^\beta} c_b(z)dz = c_b^\alpha \int_{z^\alpha}^{z^a} dz + c_b^\beta \int_{z^b}^{z^\beta} dz = L^\alpha c_b^\alpha + L^\beta c_b^\beta, \tag{2.102}$$

where superscript I is used to indicate the quantity in the surface region. The surface density of solute i is

$$\Gamma_i^H = \int_{z^\alpha}^{z^\beta} c_i(z)dz - c_i^\alpha \int_{z^\alpha}^{z^a} dz - c_i^\beta \int_{z^b}^{z^\beta} dz \tag{2.103}$$

and the real density of solute i in the surface region is

$$\Gamma_i^I = \int_{z^\alpha}^{z^\beta} c_i(z)dz = \Gamma_i^H + c_i^\alpha \int_{z^\alpha}^{z^a} dz + c_i^\beta \int_{z^b}^{z^\beta} dz = \Gamma_i^H + L^\alpha c_i^\alpha + L^\beta c_i^\beta. \tag{2.104}$$

We next consider the local thermodynamic quantity of the extensive property Y. Referring to Equation 2.15, the surface density of Y may be written as

$$y^H = \int_{z^\alpha}^{z^\beta} y(z)dz - y^\alpha \int_{z^\alpha}^{z^a} dz - y^\beta \int_{z^b}^{z^\beta} dz. \tag{2.105}$$

According to the definition of partial molar quantities, the thermodynamic quantity per unit volume in a homogeneous bulk region is given by

$$y = \frac{Y}{V} = c_a \bar{y}_a + c_b \bar{y}_b + c_i \bar{y}_i. \tag{2.106}$$

The corresponding local partial molar properties will be

$$y(z) = c_a(z)\bar{y}_a(z) + c_b(z)\bar{y}_b(z) + c_i(z)\bar{y}_i(z). \tag{2.107}$$

It follows from Equations 2.105 through 2.107 that

$$
\begin{aligned}
y^H &= \left[\int_{z^\alpha}^{z^\beta} c_a(z)\bar{y}_a(z)dz - c_a^\alpha \bar{y}_a^\alpha \int_{z^\alpha}^{z^a} dz - c_a^\beta \bar{y}_a^\beta \int_{z^b}^{z^\beta} dz \right] \\
&+ \left[\int_{z^\alpha}^{z^\beta} c_b(z)\bar{y}_b(z)dz - c_b^\alpha \bar{y}_b^\alpha \int_{z^\alpha}^{z^a} dz - c_b^\beta \bar{y}_b^\beta \int_{z^b}^{z^\beta} dz \right] \\
&+ \left[\int_{z^\alpha}^{z^\beta} c_i(z)\bar{y}_i(z)dz - c_i^\alpha \bar{y}_i^\alpha \int_{z^\alpha}^{z^a} dz - c_i^\beta \bar{y}_i^\beta \int_{z^b}^{z^\beta} dz \right]. \tag{2.108}
\end{aligned}
$$

Figure 2.2 may help to visualize the integrals in this equation. Suppose that the abscissa represents $c_j(z)y_j(z)$ instead of $c_j(z)$; then the first term in each square bracket represents the region on the left side of the line between the planes z^α and z^β. This area can be written as $\Gamma_j^I \langle \bar{y}_j^I \rangle$; here the symbol $\langle \bar{y}_j^I \rangle$ is used to represent the mean partial molar property in the surface region. The second and third terms in each square bracket will be found as the contributions of the magnitude of $\bar{y}_j$ corresponding to bulk phases. Then the first and second terms in Equation 2.108 can be written as

$$\Gamma_j^I \Delta y_j = \Gamma_j^I \langle \bar{y}_j^I \rangle - c_j^\alpha \bar{y}_j^\alpha L^\alpha - c_j^\beta \bar{y}_j^\beta L^\beta, \quad j = a,b \tag{2.109}$$

Motomura (1978a) has noted that the third term is virtually equal to $\Gamma_i^H \langle y_i^H \rangle$ when the solute is surface active. Thus, we may write Equation 2.108 in the form

$$y^H = \Gamma_a^I \Delta y_a + \Gamma_b^I \Delta y_b + \Gamma_i^H \langle \bar{y}_i^H \rangle. \tag{2.110}$$

Combined with Equations 2.93 and 2.94,

$$\Delta s = \Gamma_a^I \Delta s_a + \Gamma_b^I \Delta s_b + \Gamma_i^H \left(\left\langle \overline{s}_i^H \right\rangle - \overline{s}_i^\alpha \right) \tag{2.111}$$

and

$$\Delta v = \Gamma_a^I \Delta v_a + \Gamma_b^I \Delta v_b + \Gamma_i^H \left(\left\langle \overline{v}_i^H \right\rangle - \overline{v}_i^\alpha \right). \tag{2.112}$$

For the thermodynamic quantity y,

$$\Delta y = \Gamma_a^I \Delta y_a + \Gamma_b^I \Delta y_b + \Gamma_i^H \left(\left\langle \overline{y}_i^H \right\rangle - \overline{y}_i^\alpha \right). \tag{2.113}$$

In order to facilitate the understanding of these quantities, it is useful to consider the system consisting of immiscible solvents (cf. Figure 2.3). When the magnitudes of both c_a^β and c_b^α are sufficiently less than that of c_a^α or c_b^β, we may write

$$\Delta s = \Gamma_a^I \left(\left\langle \overline{s}_a^I \right\rangle - \overline{s}_a^\alpha \right) + \Gamma_b^I \left(\left\langle \overline{s}_b^I \right\rangle - \overline{s}_b^\beta \right) + \Gamma_i^H \left(\left\langle \overline{s}_i^H \right\rangle - \overline{s}_i^\alpha \right) \tag{2.114}$$

and

$$\Delta v = \Gamma_a^I \left(\left\langle \overline{v}_a^I \right\rangle - \overline{v}_a^\alpha \right) + \Gamma_b^I \left(\left\langle \overline{v}_b^I \right\rangle - \overline{v}_b^\beta \right) + \Gamma_i^H \left(\left\langle \overline{v}_i^H \right\rangle - \overline{v}_i^\alpha \right). \tag{2.115}$$

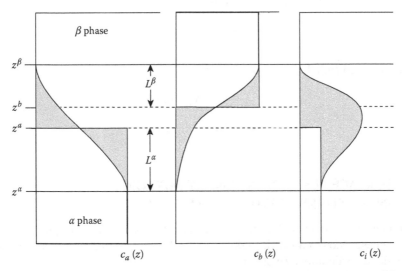

FIGURE 2.3 Schematic distributions of the constituents a, b, and i under the condition that solvents a and b are miscible only in the surface region and that the solute i does not dissolve in phase β.

The right side of these equations represents the change in the entropy and volume observed on the formation of the interface from given bulk phases. These are called the entropy and volume of interface formation, respectively (Motomura 1978, Eq. 74).

Referring to Figure 2.3 we can write

$$\Gamma_i^H = \Gamma_i^I - \int_{z^\alpha}^{z^a} c_i^\alpha dz. \tag{2.116}$$

Using the definition $\Delta y = y^H - \Gamma_i^{H-\alpha} y_i$ and Equation 2.108, the Δs and Δv can be written

$$\Delta s = \Gamma_a^I \left(\left\langle \overline{s}_a^I \right\rangle - \overline{s}_a^\alpha \right) + \Gamma_b^I \left(\left\langle \overline{s}_b^I \right\rangle - \overline{s}_b^\beta \right) + \Gamma_i^I \left(\left\langle \overline{s}_i^I \right\rangle - \overline{s}_i^\alpha \right) \tag{2.117}$$

and

$$\Delta v = \Gamma_a^I \left(\left\langle \overline{v}_a^I \right\rangle - \overline{v}_a^\alpha \right) + \Gamma_b^I \left(\left\langle \overline{v}_b^I \right\rangle - \overline{v}_b^\beta \right) + \Gamma_i^I \left(\left\langle \overline{v}_i^I \right\rangle - \overline{v}_i^\alpha \right). \tag{2.118}$$

When the system is in an equilibrium state, the Gibbs free energy of interface formation is given by

$$\Delta g = \Gamma_a^I \left(\mu_a^I - \mu_a^\alpha \right) + \Gamma_b^I \left(\mu_b^I - \mu_b^\beta \right) + \Gamma_i^I \left(\mu_i^I - \mu_i^\alpha \right) = 0. \tag{2.119}$$

Making use of the definitions $G = F + pV - \gamma A$,

$$\Delta f = \gamma - p\Delta v. \tag{2.120}$$

The energy and enthalpy of adsorption are

$$\Delta u = \gamma + T\Delta s - p\Delta v, \tag{2.121}$$

$$\Delta h = T\Delta s. \tag{2.122}$$

2.9   SURFACE TENSION OF BINARY MIXTURES IN EQUILIBRIUM WITH A WATER PHASE

Let us consider a system of binary mixtures of solvents a_1 and a_2 in equilibrium with a water phase (Motomura et al. 1988). If we modify Equation 2.18 to meet the condition under consideration, we obtain

$$d\gamma = -s^\sigma dT + v^\sigma dp - \Gamma_{a1}^\sigma d\mu_{a1} - \Gamma_{a2}^\sigma d\mu_{a2} - \Gamma_b^\sigma d\mu_b. \tag{2.123}$$

Since the number of degrees of freedom is three, two of the five independent variables should be deleted by means of two dividing surfaces. Motomura prefers to use two dividing surfaces such as

$$\Gamma_{a1} + \Gamma_{a2} = \frac{n^{\sigma}_{a1} + n^{\sigma}_{a2}}{A} = \frac{n_{a1} + n_{a2}}{A} - l^{\alpha}\left(c^{\alpha}_{a1} + c^{\alpha}_{a2}\right) - l^{\beta}\left(c^{\beta}_{a1} + c^{\beta}_{a2}\right) = 0 \quad (2.124)$$

and

$$\Gamma_b = \frac{n^{\sigma}_b}{A} = \frac{n_b}{A} - l^{\alpha}c^{\alpha}_b - l^{\beta}c^{\beta}_b = 0. \quad (2.125)$$

Under these conditions, Equation 2.123 reduces to

$$d\gamma = -s^M dT + v^M dp - \Gamma^M_{a2}d(\mu_{a2} - \mu_{a1}), \quad (2.126)$$

where superscript M is used to indicate the definition of dividing surfaces. The chemical potentials of the solvents a_1 and a_2 of the binary mixtures are a function of T, p, and x_{a2}, and the total differential of them is written in the form

$$d\mu^{\alpha}_{a1} = -\bar{s}^{\alpha}_{a1}dT + \bar{v}^{\alpha}_{a1}dp + \left(\frac{\partial\mu^{\alpha}_{a1}}{\partial x^{\alpha}_{a2}}\right)dx^{\alpha}_{a2} \quad (2.127)$$

and

$$d\mu^{\alpha}_{a2} = -\bar{s}^{\alpha}_{a2}dT + \bar{v}^{\alpha}_{a2}dp + \left(\frac{\partial\mu^{\alpha}_{a2}}{\partial x^{\alpha}_{a2}}\right)dx^{\alpha}_{a2}. \quad (2.128)$$

By substitution of these in Equation 2.126, we have

$$d\gamma = -\left(s^M - \Gamma^M_{a1}s^{\alpha}_{a1} - \Gamma^M_{a2}s^{\alpha}_{a2}\right)dT + \left(v^M - \Gamma^M_{a1}v^{\alpha}_{a1} - \Gamma^M_{a2}v^{\alpha}_{a2}\right)dp$$

$$- \Gamma^M_{a1}\left(\frac{\partial\mu^{\alpha}_{a1}}{\partial x^{\alpha}_2}\right)dx^{\alpha}_{a2} - \Gamma^M_{a2}\left(\frac{\partial\mu^{\alpha}_{a2}}{\partial x^{\alpha}_2}\right)dx^{\alpha}_{a2}. \quad (2.129)$$

The Gibbs–Duhem equation of the binary mixture at constant temperature and pressure is

$$x_{a1}\frac{d\mu_{a1}}{dx_{a2}} + x_{a2}\frac{d\mu_{a2}}{dx_{a2}} = 0.$$

With the aid of this relation, Equation 2.129 is rewritten in the form

$$d\gamma = -\left(s^M - \Gamma^M_{a1}s^{\alpha}_{a1} - \Gamma^M_{a2}s^{\alpha}_{a2}\right)dT + \left(v^M - \Gamma^M_{a1}v^{\alpha}_{a1} - \Gamma^M_{a2}v^{\alpha}_{a2}\right)dp - \frac{\Gamma^M_{a2}}{x_{a1}}\left(\frac{\partial\mu^{\alpha}_{a2}}{\partial x^{\alpha}_{a2}}\right)dx^{\alpha}_{a2}.$$

$$(2.130)$$

2.10   SURFACE TENSION OF MIXED SOLUTE SYSTEM

Let us consider the adsorption mixed solution, where the system is a three-component (solvent a, solutes 1 and 2 in phase α, and solvent b in phase β) two-phase system (α and β). For this system, Equation 2.70 can be rewritten as

$$d\gamma = -s^H dT + v^H dp - \Gamma_1^H d\mu_1 - \Gamma_2^H d\mu_2 \qquad (2.131)$$

When solvents a and b are practically immiscible, the chemical potential is described as a function of T, p, x_1^α, and x_2^α:

$$d\mu_1^\alpha = -\bar{s}_1^\alpha dT + \bar{v}_1^\alpha dp + \left(\frac{\partial \mu_1}{\partial x_1^\alpha}\right)_{T,p,x_2} dx_1^\alpha + \left(\frac{\partial \mu_1}{\partial x_2^\alpha}\right)_{T,p,x_1} dx_2^\alpha \qquad (2.132)$$

and

$$d\mu_2^\alpha = -\bar{s}_2^\alpha dT + \bar{v}_2^\alpha dp + \left(\frac{\partial \mu_2}{\partial x_1^\alpha}\right)_{T,p,x_2} dx_1^\alpha + \left(\frac{\partial \mu_2}{\partial x_2^\alpha}\right)_{T,p,x_1} dx_2^\alpha. \qquad (2.133)$$

Substitution of these equations into Equation 2.131 leads to basic relations; at the given pressure and concentrations,

$$(d\gamma)_{p,x_1^\alpha,x_2^\alpha} = -\left(s^H - \Gamma_1^H s_1^\alpha - \Gamma_2^H s_2^\alpha\right) dT = -\Delta s\, dT; \qquad (2.134)$$

and at the given temperature and concentrations,

$$(d\gamma)_{T,x_1^\alpha,x_2^\alpha} = \left(v^H - \Gamma_1^H v_1^\alpha - \Gamma_2^H v_2^\alpha\right) dp = \Delta v\, dp; \qquad (2.135)$$

at the given temperature and pressure,

$$(d\gamma)_{T,p} = -\left[\Gamma_1^H \left(\frac{\partial \mu_1}{\partial x_1^\alpha}\right)_{T,p,x_2^\alpha} + \Gamma_2^H \left(\frac{\partial \mu_2}{\partial x_1^\alpha}\right)_{T,p,x_2^\alpha}\right] dx_1^\alpha$$

$$-\left[\Gamma_1^H \left(\frac{\partial \mu_1}{\partial x_2^\alpha}\right)_{T,p,x_1^\alpha} + \Gamma_2^H \left(\frac{\partial \mu_2}{\partial x_2^\alpha}\right)_{T,p,x_1^\alpha}\right] dx_2^\alpha. \qquad (2.136)$$

If we assume that phase α is an ideal solution, the chemical potential of component i is represented by

$$\mu_i = \mu_i^{0,\alpha}(T,p) + RT \ln x_i^\alpha. \qquad (2.137)$$

The derivatives with respect to concentration are

$$\frac{\partial \mu_i^\alpha}{\partial x_i^\alpha} = \frac{RT}{x_i^\alpha} \tag{2.138}$$

and

$$\frac{\partial \mu_i^\alpha}{\partial x_j^\alpha} = 0, \quad j \neq i. \tag{2.139}$$

Thus, for the ideal dilute solution, Equation 2.136 reduces to

$$(d\gamma)_{T,p} = -\Gamma_1^H \frac{RT}{x_1^\alpha} dx_1^\alpha - \Gamma_2^H \frac{RT}{x_2^\alpha} dx_2^\alpha. \tag{2.140}$$

When we consider the mixture under fixed temperature and pressure, it is convenient to use two variables x and X defined as follows instead of x_1^α and x_2^α:

$$x_t^\alpha = x_1^\alpha + x_2^\alpha \tag{2.141}$$

and

$$X_2 = \frac{x_2^\alpha}{x_1^\alpha + x_2^\alpha}. \tag{2.142}$$

From these equations, we have two simultaneous equations

$$dx_t^\alpha = dx_1^\alpha + dx_2^\alpha \tag{2.143}$$

and

$$dX = -\frac{X_2}{x_t^\alpha} dx_1^\alpha + \frac{1-X_2}{x_t^\alpha} dx_2^\alpha. \tag{2.144}$$

From these equations,

$$dx_1^\alpha = (1-X_2)dx_t^\alpha - x_t^\alpha dX_2 \tag{2.145}$$

and

$$dx_2^\alpha = X_2 dx_t^\alpha + x_t^\alpha dX_2. \tag{2.146}$$

Assuming that the solution is a dilute ideal solution,

$$(d\gamma)_{T,p} = -RT \frac{\Gamma_t^H}{x_t^\alpha} dx_t^\alpha + RT \frac{\Gamma_t^H}{X_1 X_2} \left(X_2 - X_2^H \right) dX_2. \tag{2.147}$$

Here, Γ_t^H and X_2^H are defined as

$$\Gamma_t^H = \Gamma_1^H + \Gamma_2^H \tag{2.148}$$

and

$$X_2^H = \frac{\Gamma_2^H}{\Gamma_1^H + \Gamma_2^H}. \tag{2.149}$$

Rearranging Equation 2.147 leads to

$$dx_t^\alpha = \frac{x_t^\alpha}{RT\Gamma_t^H}(d\gamma)_{T,p} + \frac{x_t^\alpha}{X_1 X_2}\left(X_2 - X_2^H\right)dX_2. \tag{2.150}$$

If we plot x_t^α against X_2 at the given value of γ, the difference in composition between the bulk solution and the surface region can be calculated from the slope of the curve.

2.11   CONCENTRATION VARIABLES AND CHEMICAL POTENTIALS OF ELECTROLYTE SOLUTIONS

As in the case of acid–base reactions, many properties in the chemistry discipline involve evaluation of the concentration of individual ions. Chemical equilibriums in solutions are commonly described in terms of the concentrations of ions. When a strong electrolyte $M_{\nu_+}^{z_+} X_{\nu_-}^{z_-}$ is dissolved in water, it dissociates completely into ν_+ cations of charge number z_+ and ν_- anions of charge number z_-:

$$M_{\nu_+}^{z_+} X_{\nu_-}^{z_-} \to \nu_+ M^{z_+} + \nu_- X^{z_-}. \tag{2.151}$$

The properties of a solution will be determined by a number of ionic constituents and their concentrations.

The molality, in moles per 1000 g of solvent, of cations m_+ and that of anions m_- are

$$m_+ = \nu_+ m_i \tag{2.152}$$

and

$$m_- = \nu_- m_i, \tag{2.153}$$

where m_i is the molality of the electrolyte i defined by an empirical formula. The chemical potential of an ion can be written in terms of individual molality of ions in the form

$$\mu_+ = \mu_+^0 + RT \ln f_+ m_+ \tag{2.154}$$

and

$$\mu_- = \mu_-^0 + RT \ln f_- m_-, \tag{2.155}$$

where f is the activity coefficient of ions. If we introduce partial molar quantity $\bar{y}_i$ for an electrolyte i,

$$\bar{y}_i = v_+\bar{y}_+ + v_-\bar{y}_-. \tag{2.156}$$

Then, the chemical potential of a component i in a solution may be written as

$$\mu_i - \mu_i^0 = RT \ln\left[(f_+m_+)^{v_+}(f_+m_+)^{v_+}\right] = RT \ln(f_\pm m_\pm)^{(v_++v_-)}, \tag{2.157}$$

where

$$f_\pm = \left(f_+^{v_+} f_-^{v_-}\right)^{1/(v_++v_-)} \tag{2.158}$$

and

$$m_\pm = \left(m_+^{v_+} m_-^{v_-}\right)^{1/(v_++v_-)} \tag{2.159}$$

and formally $\mu_\pm$ is defined as

$$\mu_\pm = \frac{\mu_i}{v_+ + v_-} = \mu_\pm^0 + RT \ln(f_\pm m_\pm). \tag{2.160}$$

In a solution of an electrolyte, electrical neutrality requires that

$$z_+n_+ + z_-n_- = z_+m_+ + z_-m_- = 0. \tag{2.161}$$

It is also presupposed that the electroneutrality would apply to ions in the surface region as well. Hence

$$z_+\Gamma_+^H + z_-\Gamma_-^H = 0. \tag{2.162}$$

We also assume that the electrolyte i dissociates completely in the surface. Then

$$\Gamma_i^H = \frac{\Gamma_+^H}{v_+} = \frac{\Gamma_-^H}{v_-}, \tag{2.163}$$

where Γ_i^H is the surface density of the electrolyte i defined by formula weight.

2.12   SURFACE TENSION OF ELECTROLYTE SOLUTIONS

Let us consider a system in which the aqueous solution of a strong electrolyte is in equilibrium with an air phase, where the mutual solubility of the aqueous solution and air can be negligible. Under the condition that ions may not move under an external electric field, the ordinary chemical potentials of ions can be used instead of the electrochemical potentials. The variation of surface tension is given by

$$d\gamma = -s^H dT + v^H dp - \Gamma_+^H d\mu_+ - \Gamma_-^H d\mu_-. \tag{2.164}$$

Since air and water are immiscible and the electrolytes are insoluble in the air phase, the chemical potential of the ion k in the water phase is described as a function of temperature, pressure, and concentration of the electrolyte i. Hence,

$$d\mu_k = -\bar{s}_k dT + \bar{v}_k dp + \left(\frac{d\mu_k}{dm_i}\right)_{T,p} dm_i. \tag{2.165}$$

Substituting Equation 2.165 into (2.164) results in

$$dy = -\left(s^H - \Gamma_+^H \bar{s}_+ - \Gamma_-^H \bar{s}_-\right) dT$$
$$+ \left(v^H - \Gamma_+^H \bar{v}_+ - \Gamma_-^H \bar{v}_-\right) dp - \left[\Gamma_+^H \left(\frac{d\mu_+}{dm_i}\right) + \Gamma_-^H \left(\frac{d\mu_-}{dm_i}\right)\right] dm_i. \tag{2.166}$$

It is convenient to express the quantities in terms of that of an electrolyte instead of ions. By making use of Equations 2.163 and 2.156, Equation 2.166 is rewritten in the form

$$dy = -\left(s^H - \Gamma_i^H \bar{s}_i\right) dT + \left(v^H - \Gamma_i^H \bar{v}_i\right) dp - (v_+ + v_-)\Gamma_i^H \left(\frac{d\mu_\pm}{dm_i}\right) dm_i. \tag{2.167}$$

This equation has the same form as Equation 2.92, though it may seem somewhat inconvenient. By using the notation of Equation 2.92,

$$dy = -\Delta s \, dT + \Delta v \, dp - (v_+ + v_-)\Gamma_i^H \left(\frac{d\mu_\pm}{dm_i}\right) dm_i. \tag{2.168}$$

In order to calculate the magnitude of Γ_i^H from the surface tension–concentration relation data and from the earlier equation, we need to write $d\mu_\pm$ in terms of m_i. However, the definition of $\mu_\pm$ gives some notion of the complexity of the form of the chemical potentials and concentrations. For the electrolyte solutions, it is important to point out two consequences of the theory. First, the chemical potential is defined based on the number of moles of an ion as a constituent instead of an electrolyte as a component, because dissolved ions determine the properties of solutions. Second, the concentration of an electrolyte is commonly used as an independent variable, because the concentration of the individual ionic species depends upon the rule of electroneutrality. According to the Gibbs phase rule, y should be a function of three variables T, p, and the concentrations of the components. Thus, using the chain rule of differential calculus, we obtain

$$d\mu_k = \left(\frac{d\mu_k}{dm_k}\right)_{T,p} \left(\frac{dm_k}{dm_i}\right) dm_i = \frac{v_k RT}{m_k}\left(1 + \frac{d\ln f_k}{d\ln m_k}\right) dm_i. \tag{2.169}$$

The total differential of the surface tension can be obtained by substituting this relation into Equation 2.166 and is given by

$$d\gamma = -\Delta s\, dT + \Delta v\, dp - RT\frac{(v_+ + v_-)\Gamma_i^H}{m_i}\left(1 + \frac{d\ln f_\pm}{d\ln m_i}\right)dm_i. \tag{2.170}$$

By means of the surface tension measurements of the aqueous solution of an electrolyte i as a function of temperature, pressure, and concentration, the entropy change, volume change, and the surface excess density of i are expressed as

$$\Delta s = -\left(\frac{d\gamma}{dT}\right)_{p,m_i}, \tag{2.171}$$

$$\Delta v = \left(\frac{d\gamma}{dp}\right)_{T,m_i}, \tag{2.172}$$

and

$$\Gamma_i^H = -\frac{m_i}{RT(v_+ + v_-)}\left(\frac{d\gamma}{dm_i}\right)_{T,p}\bigg/\left(1 + \frac{d\ln f_\pm}{d\ln m_i}\right). \tag{2.173}$$

Next, let us consider the quasithermodynamic expression of the thermodynamic quantities of adsorption of an electrolyte i. The thermodynamic quantity per unit volume in a homogeneous bulk phase is given by

$$y = \frac{Y}{V} = c_a\bar{y}_a + c_b\bar{y}_b + c_+\bar{y}_+ + c_-\bar{y}_-. \tag{2.174}$$

The quantities c_+ and c_- are equal to v_+c_i and v_-c_i, respectively. Then with the aid of Equation 2.156, Equation 2.174 is reduced to

$$y = c_a\bar{y}_a + c_b\bar{y}_b + c_i\bar{y}_i. \tag{2.175}$$

This equation is identical to Equation 2.106 in form. Then, we can write for the aqueous solution of electrolyte i

$$\Delta y = \Gamma_a^I\Delta y_a + \Gamma_b^I\Delta y_b + \Gamma_i^H\left(\langle\bar{y}_i^H\rangle - \bar{y}_i^\alpha\right). \tag{2.176}$$

In general, ions are soluble only in phase α:

$$\Delta y = \Gamma_a^I\left(\langle\bar{y}_a^I\rangle - \bar{y}_a^\alpha\right) + \Gamma_b^I\left(\langle\bar{y}_b^I\rangle - \bar{y}_b^\beta\right) + \Gamma_i^I\left(\langle\bar{y}_i^I\rangle - \bar{y}_i^\alpha\right). \tag{2.177}$$

2.13 TOTAL NUMBER OF MOLES OF IONS AS A CONCENTRATION VARIABLE

In general, the number of moles of electrolytes based on its empirical formula is used as one of the independent variables, but the use of the number of moles of ions will be still attractive. Motomura et al. (1984) has introduced new concentration variables $\hat{m}_i$ based on the number of moles of ions. The $\hat{m}_i$ of a single electrolyte solution that dissociates completely into ν_+ cations of charge number z_+ and ν_- anions of charge number z_- is given by

$$m = m_+ + m_- = (\nu_+ + \nu_-)m_i. \tag{2.178}$$

From the relation

$$\frac{m_+}{\nu_+} = \frac{m_-}{\nu_-}, \tag{2.179}$$

which follows from the condition of electroneutrality, we may write

$$dm = \frac{\nu_+ + \nu_-}{\nu_+} dm_+ = \frac{\nu_+ + \nu_-}{\nu_-} dm_-. \tag{2.180}$$

In a similar manner, the adsorption of ions is defined as

$$\Gamma^H = \Gamma_+^H + \Gamma_-^H = (\nu_+ + \nu_-)\Gamma_i^H. \tag{2.181}$$

Then, Equation 2.164 can be reduced to

$$d\gamma = -\Delta s dT + \Delta v dp - RT \frac{\Gamma^H}{m}\left(1 + \frac{d \ln f_\pm}{d \ln m}\right)dm, \tag{2.182}$$

which has the same form as derived for nonelectrolyte solutions.

REFERENCES

Gibbs, J. W. 1875–1878. Influences of surfaces of discontinuity upon the equilibrium of heterogeneous masses. Theory of capillarity. In *Collected Works*, Vol. 1, pp. 219–331, Ox Bow Press, Woodbridge, 1993.

Hill, T. L. 1952. Statistical thermodynamics of the transition region between two phases. I. Thermodynamics and quasi-thermodynamics. *J. Phys. Chem.* 56: 526–531.

Kirkwood, J. G. and I. Oppenheim. 1961. *Chemical Thermodynamics*. McGraw-Hill, New York.

Motomura, K. 1978. Thermodynamic studies on adsorption at interfaces I. General formulation. *J. Colloid Interface Sci.* 64: 348–355.

Motomura, K. 1980. Thermodynamics of interfacial monolayers. *Adv. Colloid Interface Sci.* 12: 1–42.

Motomura, K., N. Ando, H. Matsuki, and M. Aratono. 1990. Thermodynamic studies on adsorption at interfaces VII. Adsorption and micelle formation of binary surfactant mixtures. *J. Colloid Interface Sci.* 139: 188–197.

Motomura, K., M. Aratono, N. Matubayasi, and R. Matuura. 1978a. Thermodynamic stud-
ies on adsorption at interfaces III. Sodium dodecyl sulfate at water/hexane interface.
J. Colloid Interface Sci. 67: 247–254.
Motomura, K., S. Iwanaga, Y. Hayami, S. Uryu, and R. Matuura. 1980. Thermodynamic stud-
ies on adsorption at interfaces IV. Dodecylammonium chloride at water/air interface.
J. Colloid Interface Sci. 80: 32–38.
Motomura, K., S. Iwanaga, M. Yamanaka, M. Aratono, and R. Matuura. 1982. Thermodynamic
studies on adsorption at interfaces V. Adsorption from micellar solution. *J. Colloid
Interface Sci.* 86: 151–157.
Motomura, K., H. Iyota, N. Ikeda, and M. Aratono. 1988. Thermodynamic studies on adsorp-
tion at interfaces VI. Interface between cyclohexane – benzene mixture and water.
J. Colloid Interface Sci. 123: 26–36.
Motomura, K., N. Matubayasi, M. Aratono, and R. Matuura. 1978b. Thermodynamic studies
on adsorption at interfaces II. One surface-active component system: Tetradecanol at
hexane/water interface. *J. Colloid Interface Sci.* 64: 356–361.
Motomura, K., M. Yamanaka, and M. Aratono. 1984. Thermodynamic consideration of the
mixed micelle of surfactants. *Colloid Polym. Sci.* 262: 948–955.
Ono, S. and S. Kondo. 1960. Molecular theory of surface tension in liquids. In *Handbuch der
Physic*, ed. S. Flügge, Vol. X, pp. 134–278, Springer-Verlag, Berlin, Germany.

3 Surface Tension of Pure Water at Air/Water and Oil/Water Interfaces

Surface tension measurements of solutions have been a major concern in surface science, but an understanding of the nature of the thermodynamic quantities involved in the adsorption may best be obtained by considering the fundamental nature of the pure water surface. In this chapter, a comprehensive discussion of thermodynamic quantities of the pure water surface will be made as a beginning of the systematic discussion of the solutions. Water is typical of a structured liquid with high density and high boiling temperature. The existence of many unusual physical properties of substances has been known, and the origin of the properties has been concerned with unique properties of water. Before proceeding to an outline of the water surface, it will be profitable to show the surface tension of simple liquids.

3.1 SURFACE TENSION OF ARGON, XENON, AND NITROGEN

The surface tension of a simple liquid in contact with its vapor is usually given as a function of temperature because a one-component two-phase system is univariant. The lines drawn in Figure 3.1 are values of liquid argon, xenon, and nitrogen over their liquid ranges from triple point to critical point (Stansfield 1958, Smith et al. 1967). Their values decrease steeply with rising temperature and approach to zero surface tension at the critical temperature. Equation 2.40 suggests that the superficial entropy of these simple liquids is positive. In a univariant system, the pressure of the system changes along with the temperature change, and the variation of surface tension with pressure can be drawn directly from Figure 3.1. The surface tension decreases with increasing pressure and asymptotically approaches zero at the critical point (Figure 3.2). The negative slope of this curve shows that the superficial volume has negative values. We see that molar volume in the surface region is smaller than that in the gaseous phase. This result may seem to be inconsistent with the result for the entropy change. Equation 2.39 suggests that the value is determined by the sum of the contributions in volume change from the gaseous phase to the surface and from the liquid phase to the surface. The adsorption from the gaseous phase may be accompanied by large negative volume change, while the adsorption from the liquid phase may be small. The surface region may have a liquid-like structure.

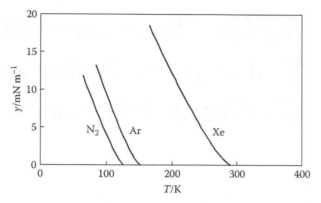

FIGURE 3.1 Variation of surface tension of liquid argon, xenon, and nitrogen in equilibrium with their vapor as a function of temperature between triple point and critical point. (Data from Stansfield, D., *Proc. Phys. Soc.*, 72, 854, 1958; Smith, B.L. et al., *J. Chem. Phys.*, 47, 1148, 1967.)

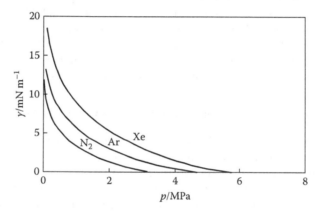

FIGURE 3.2 Surface tension versus pressure curves of liquid argon, xenon, and nitrogen in equilibrium with their vapor. (Data from Stansfield, D., *Proc. Phys. Soc.*, 72, 854, 1958; Smith, B.L. et al., *J. Chem. Phys.*, 47, 1148, 1967.)

3.2 SURFACE TENSION OF VAPOR/WATER INTERFACE

Vargafik et al. (1983) have made a detailed critical survey of the methods and the numerical values of pure water, and they provided the empirical equation that they write in the form

$$\gamma = B\left[\frac{T_c - T}{T_c}\right]^{\mu}\left[1 + b\left(\frac{T_c - T}{T_c}\right)\right], \tag{3.1}$$

where
    $T_c = 647.15$ K
    $B = 235.8$ mN m^{-1}
    $b = -0.625$
    $\mu = 1.256$

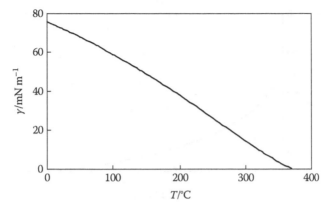

FIGURE 3.3   Surface tension–temperature curve of water in equilibrium with its vapor as a function of temperature between triple point and critical point. (Data from Vargafik, B.N. et al., *J. Phys. Chem. Ref. Data*, 12, 817, 1983.)

Figure 3.3 represents the surface tension of water between the triple point and the critical point evaluated by the empirical Equation 3.1. It decreases progressively with temperature from 75.64 mN m^{-1} at 0.01°C to 0.45 mN m^{-1} at 370°C and approaches the point of zero surface tension asymptotically to the critical point located at 374°C and 21.8 MPa. Differentiation of γ with respect to T gives the superficial entropy as shown in Figure 3.4. In comparison to Figure 3.1, water has a particularly high value of surface tension, and the experimental γ–T curve for water clearly concaves downward while the curves for simple liquids concave upward. It is important to note that the superficial entropy of pure liquid water has positive values. Δs increases with increasing temperature almost linearly, passes through a maximum, and turns downward steeply as the temperature approaches closer to the critical point.

It is also a matter of importance to consider the magnitude of Δv for the pure water surface in equilibrium with its vapor. Since measurements are made along the pressure–temperature coexistence curve, the surface tension values calculated

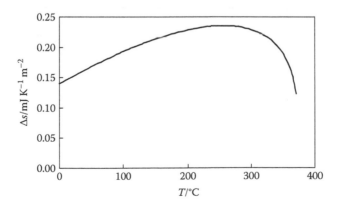

FIGURE 3.4   Entropy of surface formation of water in equilibrium with its vapor.

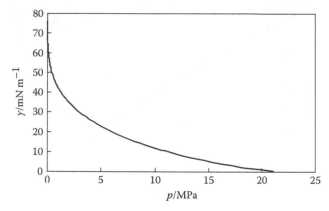

FIGURE 3.5 Variation of surface tension of water with pressure. (Data from Massoudi, R. and King, Jr., A.D., *J. Phys. Chem.*, 78, 2262, 1974.)

from Equation 3.1 can be easily plotted against the pressure, as shown in Figure 3.5. The vapor pressure of water is taken from the work of Scott and Osborn (1979). The decreases of the surface tension with increasing pressure will agree, in general, with the empirical relation that the surface tension is proportional to the density difference between liquid and vapor phases, which is known as parachor (Defay et al. 1966, p. 156). In Figure 3.6, $-\Delta v$ obtained from γ versus p data is plotted against $1/p$ from 0.1 MPa to the critical point. The negative Δv values are observed over the entire range of pressure. The pressure dependence of Δv is such that Δv changes rapidly at low pressures and approaches asymptotically to zero with increasing pressure, but it changes almost linearly with respect to $1/p$. It is important to note that the earlier results are not in accord with the results for the evaporation of water suggested by Clapeyron's equation. The entropy of evaporation of water is proportional to the volume of evaporation, but the positive Δs is accompanied by a negative change in Δv. Δs is the surface excess entropy defined by the two dividing surfaces such that $\Gamma_{water} = 0$ and $v^{\sigma} = 0$, and Δv is defined by the surfaces such that $\Gamma_{water} = 0$ and $\Delta s = 0$.

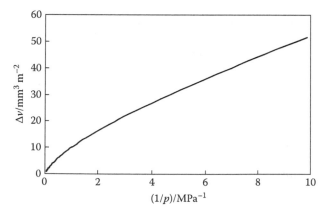

FIGURE 3.6 Volume of surface formation of water in equilibrium with its vapor.

3.3   SURFACE TENSION OF AIR/WATER INTERFACE

In general, the surface tension measurement of a two-component two-phase system is simply made as a function of temperature and pressure but neither the chemical potentials of the water nor the air. We will use Equation 2.58 for the consideration of the system:

$$d\gamma = -\Delta s dT + \Delta v dp. \tag{2.58}$$

Many precise measurements of the surface tension of pure water as a function of temperature at the atmospheric pressure have been reported, and the results of these measurements follow a monotonic decreasing relationship of the form

$$\gamma = a + bT + cT^2. \tag{3.2}$$

Using the values of coefficients reported in some literatures, numerical values of the surface tension are compared in Table 3.1, and some of them are plotted in Figure 3.7. It is clear from the results that these values obtained independently in several of the measurements show considerable discrepancy of roughly 1 mN m^{-1} even in the ordinary temperature ranges around room temperature.

TABLE 3.1

Surface Tension of Pure Water γ/mN m^{-1} as a Function of Temperature

Temperature (°C)									$d\gamma/dT$	
15.0	**17.50**	**20.0**	**22.5**	**25.0**	**27.5**	**30.0**	**32.5**	**35.0**	**(mN m^{-1} kg^{-1} mol)**	
74.10	73.73	73.36	72.98	72.60	72.21	71.82	71.43	71.03	−0.154	Kayser (1976)
73.81	73.43	73.06	72.68	72.29	71.90	71.50	71.10	70.69	−0.156	Gittens (1969)
73.79	73.41	73.03	72.64	72.25	71.86	71.46	71.07	70.67	−0.156	Gittens (1969)
73.50	73.13	72.74	72.36	71.97	71.58	71.18	70.78	70.38	−0.156	Teitel'baum et al. (1951)
73.38	72.99	72.59	72.19	71.79	71.39	70.99	70.58	70.16	−0.161	Moser (1927)
73.52	73.15	72.78	72.41	72.03	71.65	71.27	70.88	70.49	−0.151	Cini et al. (1972)
73.51	73.14	72.76	72.38	72.00	71.61	71.22	70.83	70.43	−0.154	Cini et al. (1972)
73.52	73.14	72.75	72.36	71.96	71.55	71.14	70.73	70.31	−0.161	Johansson and Eriksson[a] (1972)
73.53	73.16	72.78	72.39	72.01	71.62	71.22	70.82	70.41	−0.156	Harkins and Brown (1919)
73.49	73.12	72.74	72.36	71.98	71.59	71.20	70.80	70.41	−0.154	Vargaftik et al. (1983)

[a]   The vales of Johansson and Eriksson were rewritten by using the value of 71.96 at 25°C.

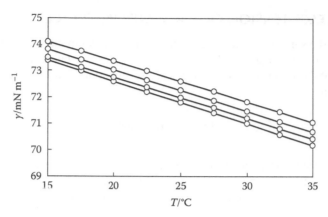

FIGURE 3.7 Surface tension of air/water interface as a function of temperature. From top to bottom, Kayser (1976), Gittens (1969), Vargaftik (1983), and Moser (1927).

Before discussing the thermodynamic aspect of these results, it will be useful to describe them briefly from an experimental viewpoint. In general, variation in the surface tension of pure water will not be a result of carelessness, although there is a tendency to account for the lower γ values by assuming some kind of contamination. In view of the discussion concerning a correlation between γ and the chemical nature of the liquid, we might expect evidence that the $d\gamma/dT$ value is specific to each liquid. Indeed, the $d\gamma/dT$ values shown in Table 3.1 show a tendency to share a common value of water. The $d\gamma/dT$ values of pure water can be used as an index of contamination better than the value of the surface tension itself. Soluble contaminants generally lead to diminution of the values, and the insoluble contaminants sometimes lead to difficulties in evaluation of the value because of the scattering of the values. In comparing the surface tension of pure water shown in Table 3.1, we are not inclined to ascribe the lower values to the contamination only. From the experimental point of view, the variation of γ itself for pure water would not be from contamination or experimental techniques but from the systematic one specific for the method and conditions employed. Since each method of measurement depends on the contact angle of the water/glass surface in some degrees, it seems probable that most part of the variation in the surface tension of water can be attributed to the method itself.

A comparison of the observed γ using capillary rise and plate methods is made in Table 3.2. The average values of the capillary rise method are 72.78 ± 0.06 and 72.12 ± 0.13 mN m^{-1} at 20°C and 25°C, respectively, while those of the plate method are 72.97 ± 0.25 and 72.13 ± 0.27 mN m^{-1}, respectively. The agreement is very satisfactory and appears to be acceptable. However, Vargaftik et al. (1983) have reviewed a large number of experimental studies of γ for water and pointed out that the capillary rise method is at present one of the most reliable methods for determining the surface tension of fluids. They have shown the standard values adopted by the International Association for the Properties of Steam, although they presented them as the surface tension between water and its vapor interface. Richards and Coombs (1915) and Richards and Carver (1921) measured the capillary rise of water in air and in the

TABLE 3.2

Surface Tension (mN m⁻¹) of Pure Water Obtained by Capillary Rise and Wilhelmy Plate Methods at 20°C and 25°C

Capillary Rise Method			Wilhelmy Plate Method		
20.0	25.0		20.0	25.0	
72.80	—	Harkins and Brown (1919)	72.00	—	Padday and Russell (1960)
72.70	—	Sugden (1921)	—	71.98	Kawanishi et al. (1970)
72.73	—	Richards and Carver (1921)	73.04	—	Taylor and Mingins (1975)
72.78	72.00	Harkins (1945)	73.36	72.6	Kayser (1976)
72.75	—	Volyak and Dokl (1950)	—	71.98	Pallas and Pethica (1983)
72.85	—	Harkins (1952)	72.80	—	Owens et al. (1987)
—	72.11	Drost-Hansen (1965)	72.94	72.13	Gaonkar and Neumann (1987)
72.88	71.14	Jasper (1972)	—	—	—
72.75	71.99	Vargafik et al. (1983)	—	—	—
—	72.04	Pallas and Pethica (1983)	—	—	—

presence of vapor without the air, and found no difference between them at 20°C. The surface tension values at the vapor/water interface of Vargaftik are shown in the ninth column of Table 3.1. It is to be noted that γ values shown in the table are measured in the presence of a variety of gaseous phases such as nitrogen, helium, and air.

Practically, in all measurements of the surface tension of the air/water interface, it is customary to select the temperature as an independent variable keeping the pressure constant. It is hard to search the literature on the pressure effect on the surface tension of an air/water interface, but the experimental data of Massoudi and King Jr. (1974, 1975), and Hough et al. (1952), who reported systematic studies on the adsorption of gases on water surface, are helpful in summarizing the general behavior of γ of the air/water interface. The γ values of nitrogen/water and oxygen/water interfaces at 25°C are given as a function of pressure by the following empirical relations:

$$\gamma = \gamma_0 - 0.0835p + 1.94 \times 10^4 p^2; \quad \text{nitrogen/water}\,(p/\text{atm}) \tag{3.3}$$

$$\gamma = \gamma_0 - 0.0779p + 1.04 \times 10^4 p^2; \quad \text{oxygen/water}\,(p/\text{atm}) \tag{3.4}$$

Values of γ of a nitrogen/water interface are drawn in Figure 3.8, which differs significantly from that of the water vapor/water shown in Figure 3.5. The change of the surface tension of the oxygen/water interface in the γ and p scales of this figure is almost the same as that of the nitrogen/water interface.

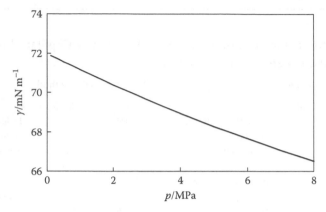

FIGURE 3.8 Surface tension of nitrogen/water interface as a function of pressure. (Data from Massoudi, R. and King, Jr., A.D., *J. Phys. Chem.*, 78, 2262, 1974.)

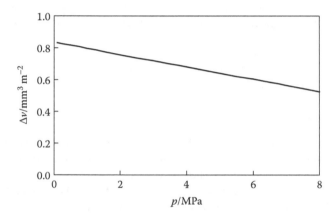

FIGURE 3.9 Volume of surface formation of water in equilibrium with nitrogen phase.

It is immediately apparent that γ versus p curves for one-component and two-component systems differ significantly, given that they both show almost the same γ versus T relationship around room temperature. The value of Δv of the nitrogen/water system is -0.83 mm^3 m^{-2} at 0.1 MPa and the negative Δv increases up to -0.52 mm^3 m^{-2} at 8 MPa (Figure 3.9), while those of the water vapor/water interface are -54 and -2 mm^3 m^{-2} at 0.1 and 8 MPa, respectively.

3.4 INTERFACIAL TENSION OF WATER IN EQUILIBRIUM WITH OIL

The determination of the effect on the interfacial tension between oil and water has its beginnings in the early history of surface science. Lynde (1906) reported that the interfacial tension of mercury/water, mercury/ether, and carbon bisulfide/water interfaces increases with the increase in the pressure while it decreases in the case of ether/water and chloroform/water interfaces. Kahlweit (1970) has suggested an explanation useful for these experimental results. He starts with Equation 1.52, which is Gibbs' fundamental equation for a two-component (oil and water) system:

$$d\gamma = -s^{\sigma} dT - \Gamma_o d\mu_o - \Gamma_w d\mu_w, \tag{3.5}$$

where subscripts o and w are used to denote oil and water, respectively. Since the number of degrees of freedom is two, μ_w is eliminated by using the convention of setting the dividing surface so that Γ_w is zero. The chemical potential of the oil in the oil phase is given as

$$d\mu_o^O = -\bar{s}_o^O dT + \bar{v}_o^O dp + \left(\frac{\partial \mu_o^O}{\partial x_o^O} \right)_{T,p} dx_o^O. \tag{3.6}$$

By substituting this expression into (3.5), we obtain

$$\left(\frac{d\gamma}{dp} \right)_T = -\Gamma_{o(w)} \left[\bar{v}_o^O + \left(\frac{\partial \mu_o^O}{\partial x_o^O} \right)_{T,p} \left(\frac{\partial x_o^O}{\partial p} \right)_T \right]. \tag{3.7}$$

For the case in which oil and water are practically immiscible, the contribution of the second term in the square bracket can be neglected. With $\bar{v}_o^O = 60.4 \, \mathrm{cm}^3 \, \mathrm{mol}^{-1}$ for CS_2, it follows from this equation that $\Gamma_{CS_2(w)} = -0.3 \, \mu\mathrm{mol} \, \mathrm{m}^{-2}$. This negative adsorption provides useful experimental information for the properties of the interface. Kahlweit suggested that if the dividing surface is chosen at which the excess concentration of water vanishes, this dividing surface is located a little toward the aqueous phase with respect to the dividing surface that would make the excess concentration of CS_2 vanish.

For a system in which equilibrium between two pure liquid phases is maintained, Equation 1.55 is given by Gibbs (Eq. 580). He noted that the coefficient of the dp terms in Equation 1.55 represents the distance between two dividing surfaces of which one would make Γ_w vanish and the other Γ_o. This setting of the two dividing surfaces is adopted by Hansen (1962). Upon introduction of these dividing surfaces, he called the coefficient of the dp terms the surface excess volume per unit area. Gibbs has alternatively noted that the coefficient of dp represents the diminution of volume due to a unit of the surface of discontinuity.

In general, the interfacial tension between two immiscible liquids at a fixed temperature increases with increasing pressure. Such an increase at constant temperature is shown in Figure 3.10 for water–hexane and water–octane systems (Matubayasi et al. 1977). The plots are straight lines, and each shows specific slopes. For the measurements made under the condition, it would be convenient to consider the pressure coefficient by Equation 2.59 instead of (3.5). Then the positive slope suggests that the formation of the oil/water interface accompanies positive Δv. This result has been obtained for a series of oil/water interfaces (Motomura et al. 1983). This positive Δv is contrasted with the negative Δv observed for a nitrogen/water interface.

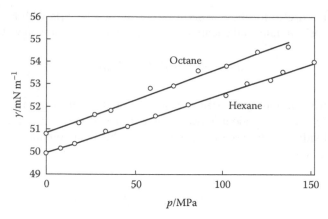

FIGURE 3.10   Variation of hexane/water and octane/water interfacial tension with pressure.

3.5   THERMODYNAMIC QUANTITIES OF SURFACE FORMATION OF PURE WATER

It is tempting to account for the thermodynamic quantities of surface formation discussed in the previous sections summarized in Table 3.3. It can be seen that observed values depend firstly on the type of the system. However, two interesting points should be noted. First, the entropy change of surface formation Δs always has positive values. Second, the magnitude of $T\Delta s$ is commonly larger than the $p\Delta v$ term.

TABLE 3.3
Thermodynamic Quantities of Surface Formation at 0.1 MPa

	T (K)	P (MPa)	γ (mN m^{-1})	Δv (mm^3 m^{-2})	Δs (mJ K^{-1} m^{-2})	$P\Delta v$	$T\Delta s$ (mJ m^{-2})	Δu
One-component								
Water	373	0.101	58.92	−53.5	0.193	−5.42	71.9	136.2
Argon	87	0.101	12.55	−23.4	0.246	−2.37	21.5	36.4
Xenon	165	0.101	18.46	−30.2	0.187	−3.06	30.8	52.4
Nitrogen	77	0.101	8.88	−19.0	0.228	−1.92	17.7	28.4
Two-component								
Air/water	298	0.101	71.96	−0.831	0.1608	−0.0831	47.9	120.0
Oil/water								
Hexane	298	0.101	50.35	0.0227	0.0916	0.00230	27.3	77.7
Octane	298	0.101	50.95	0.0278	0.0899	0.00282	26.8	77.5
Decane	298	0.101	51.41	0.0296	0.0868	0.00300	25.9	77.3
Dodecane	298	0.101	51.85	0.0311	0.0858	0.00315	25.6	77.4
Cyclohexane	298	0.101	49.86	0.0304	0.0905	0.00308	27.0	76.8
Benzene	298	0.101	34.00	0.0106	0.0604	0.00107	18.0	52.0
Butyl bromide	298	0.101	36.76	0.0173	0.0572	0.00175	17.1	53.8
Hexyl bromide	298	0.101	39.85	0.0209	0.0552	0.00212	16.5	56.3

The formation of the surface is energetically unfavorable, and this disadvantage largely depends on the heat content of surface formation.

The formation of an oil/water interface may be the simplest case, because the system consists of two condensed phases. Let us first consider the Δv values. When we are concerned with a two-phase system composed of two immiscible phases, we can write Equation 2.55 for the volume change as

$$\Delta v = \Gamma_a^I \left(\left\langle \overline{v}_a^I \right\rangle - \overline{v}_a^\alpha \right) + \Gamma_b^I \left(\left\langle \overline{v}_b^I \right\rangle - \overline{v}_b^\beta \right).$$

(2.55)

Here $\overline{v}_a^\alpha$ and $\overline{v}_b^\beta$ represent the molar volume of solvents a and b, respectively. Then, this relation suggests that Δv will be characterized by the sum of the contribution of the adsorption of a from phase α and the adsorption of b from phase β. In practice, the positive Δv of an oil/water interface suggests that mean partial molar volume of oil and/or water in the surface region will be larger than those in their bulk phases, since an enhancement of cohesion between molecules in the surface region will not be expected. Let us next consider the entropy of surface formation Δs. In general, the process of evaporation described by Clapeyron's equation is accompanied by an expansion in volume and an increment in entropy. Then we may expect that the positive Δs observed for the oil/water interface is predictable from the corresponding positive change in Δv. When two mutually immiscible components contact with each other in the surface region, there are positive Δv and Δs as shown in Table 3.3. However, we can point out a distinctive property of the surface formation. Δv for a series of linear hydrocarbons increases with increasing chain length, while Δs decreases. This can be explained if a hydrocarbon is mutually miscible with water in the surface region and their miscibility depends on the chain length. Benzene is less miscible than hydrocarbon, and Δv and Δs have smaller values than those of hexane.

For the surface region between an air/water interface, it may be convenient to consider the situation in which the water and air are practically immiscible in the bulk phase but mutually miscible in the surface region so that the mean partial molar volume of air is smaller than that which exists in the gaseous phase. In this situation, the adsorption of air into the surface region will make a negative contribution to the net volume change of surface formation. A similar consideration leads to a qualitative explanation for the remarkable negative Δv observed for the vapor–liquid equilibrium of water, provided that water vapor and liquid water are immiscible in the bulk phase with each other but freely miscible in the surface region. To obtain a quantitative understanding, however, it is convenient to consider the equation of a one-component system (2.36). It is to be noted that the negative Δv of a vapor/water interface is more remarkable than the other one-component system. The liquid-like structure of a vapor/water interface will give such a value.

REFERENCES

Defay, R., I. Prigogine, A. Bellmans, and D. H. Everett. 1966. *Surface Tension and Adsorption.* Longmans, Green & Co., London, U.K.

Gibbs, J. W. 1993. Influences of discontinuity upon the equilibrium of heterogeneous masses—Theory of capillarity. In *Collected Works*, Vol. 1. pp. 219–311, Ox Bow Press, Woodbridge, CT.

Hough, E. W., B. B. Wood Jr., and M. J. Rzasa. 1952. Adsorption at water-helium and -nitrogen interfaces at pressures to 15,000 p.s.i.a. *J. Phys. Chem.* 56: 996–999.

Kahlweit, M. 1970. On the effect of temperature and pressure on the interfacial tension between two liquid phases. *Ber. Bunsen-Ges.* 74: 636–638.

Lynde, C. J. 1906. The effect of pressure on surface tension. *Phys. Rev.* 22: 181–191.

Massoudi, R. and A. D. King Jr. 1974. Effect of pressure on the surface tension of water. Adsorption of low molecular weight gases on water at 25°. *J. Phys. Chem.* 78: 2262–2266.

Massoudi, R. and A. D. King Jr. 1975. Effect of pressure on the surface tension of aqueous solutions. Adsorption of hydrocarbon gases, carbon dioxide, and nitrous oxide on aqueous solutions of sodium chloride and tetra-n-butylammonium bromide at 25°. *J. Phys. Chem.* 79: 1670–1675.

Matubayasi, N., K. Motomura, S. Kaneshina, M. Nakamura, and R. Matuura. 1977. Effect of pressure on interfacial tension between oil and water. *Bull. Chem. Soc. Jpn.* 50: 523–524.

Moser, H. 1927. Detachment of bar. *Ann. Physik* 82: 993–1103.

Motomura, K., H. Iyota, M. Aratono, M. Yamanaka, and R. Matuura. 1983. Thermodynamic consideration of the pressure dependence of interfacial tension. *J. Colloid Interface Sci.* 93: 264–269.

Richards, T. W. and E. K. Carver. 1921. A critical study of the capillary rise method of determining surface tension, with data for water, benzene, toluene, chloroform, carbon tetrachloride, ether and dimethyl aniline. *J. Am. Chem. Soc.* 43: 827–847.

Scott, D. W. and A. G. Osborn. 1979. Representation of vapor-pressure data. *J. Phys. Chem.* 83: 2714–2723.

Smith, B. L., R. R. Gardner, and E. H. C. Parker. 1967. Surface tension and energy of liquid xenon. *J. Chem. Phys.* 47: 1148–1152.

Stansfield, D. 1958. The surface tensions of liquid argon and nitrogen. *Proc. Phys. Soc.* 72: 854–866.

SURFACE TENSION OF AIR/WATER INTERFACE

Teitel'baum, B. Ya., T. A. Gortalova, and E. E. Sidorova. 1951. *Zhur. Fiz. Khim.* 25: 911.

Vargafik, B. N., B. N. Volkov, and L. D. Volyak. 1983. International tables of the surface tension of water. *J. Phys. Chem. Ref. Data* 12: 817–820.

CAPILLARY RISE METHOD

Drost-Hansen, W. 1965. Methods of study and structural properties. *Ind. Eng. Chem.* 65(4): 18–37.

Gittens, G. J. J. 1969. Variation of surface tension of water with temperature. *Colloid Interface Sci.* 30: 406–412.

Harkins, W. D. 1945. Determination of surface and interfacial tension. In *Physical Methods of Organic Chemistry*, ed. A. Welssberger, Vol. 1, part 1. Interscience, New York.

Harkins, W. D. 1962. *The Physical Chemistry of Surface Films*. Reinfold, New York.

Harkins, W. and F. E. Brown. 1919. The determination of surface tension (free surface energy), and the weight of falling drops: The surface tension of water and benzene by the capillary height method. *J. Am. Chem. Soc.* 41: 499–524.

Jasper, J. J. 1972. The surface tension of pure liquids compounds. *J. Phys. Chem., Ref. Data* 1: 841–10009.

Pallas, N. R. and B. A. Pethica. 1983. The surface tension of water. *Colloids Surf.* 6: 221–227.

Richards, T. W. and E. K. Carver. 1921. A critical study of the capillary rise method of determining surface tension, with data for water, benzene, toluene, chloroform, carbon tetrachloride, ether and dimethyl aniline. *J. Am. Chem. Soc.* 43: 827–847.

Richards, T. W. and L. Coombs. 1915. The surface tensions of water, methyl, ethyl and iso-butyl alcohols, ethyl butyrate, benzene and toluene. *J. Am. Chem. Soc.* 37: 1656–1676.

Sugden, S. 1921. The determination of surface tension from the rise in capillary tubes. *J. Chem. Soc.* 119: 1483–1492.

Vargaftik, N. B., L. D. Volyak, and M. Volkov. 1983. Investigating the surface tension of H_2O and D_2O at near-critical temperatures. *Teploenergetika* 20(8): 80–82.

Volyak, L. and D. Dokl. 1950. Surface tension of water as a function of temperature. *Akad. Nauk. SSSR* 74: 307–310.

WILHELMY PLATE METHOD

Cini, R., G. Loglio, and A. Ficalby. 1972. Temperature dependence of the surface tension of water by the equilibrium ring method. *J. Colloid Interface Sci.* 41: 287–297.

Gaonkar, A. G. and R. D. Neumann.1987. The uncertainty in absolute values of surface tension of water. *Colloids Surf.* 27: 1–14.

Johansson, K. and J. C. Eriksson. 1972. Determination of dγ/dT for water by means of a dif-ferential technique. *J. Colloid Interface Sci.* 40: 398–405.

Kawanishi, T., T. Seimiya, and T. Sasaki. 1970. Corrections for surface tension measured by Wilhelmy method. *J. Colloid Interface Sci.* 32: 622–627.

Kayser, W. V. 1976. Temperature dependence of the surface tension of water m contact with its saturated vapor. *J. Colloid Interface Sci.* 56: 622–627.

Owens, N. F., D. S. Johnston, D. Gingell, and D. Chapman. 1987. Surface properties of a long-chain 10:12 diynoic acid monolayer at air-liquid and solid-liquid interfaces. *Thin Solid Films* 155: 255–266.

Padday, J. F. and D. R. Russell. 1960. The measurement of the surface tension of pure liquids and solutions. *J. Colloid Sci.* 15: 503–511.

Pallas, N. R. and B. A. Pethica. 1983. The surface tension of water. *Colloids Surf.* 6: 221–227.

Taylor, J. A. G. and J. Mingins. 1975. Properties of the non-polar oil/water interface. Part 1—Procedures for the accurate measurement of the interfacial pressure of an insoluble monolayer. *J. Chem. Soc. Faraday Trans. I* 71: 1161–1171.

DROP VOLUME METHOD

Gittens, G. J. 1969. Variation of surface tension of water with temperature. *J. Colloid Interface Sci.* 30: 406–412.

Harkins, W. and F. E. Brown. 1919. The determination of surface tension (free surface energy), and the weight of falling drops: The surface tension of water and benzene by the capil-lary height method. *J. Am. Chem. Soc.* 41: 499–524.

4 Surface Tension of Solutions

In the preceding chapter, the surface tension and the related thermodynamic quantities of two-component two-phase systems were discussed in detail. In this chapter, we will consider the thermodynamic quantities associated with the adsorption and desorption of solutes. When we carry out the surface tension measurement for solutions, we obtain information about surface excess of solute, which we call adsorption. The adsorption is the thermodynamic quantity defined by the adsorption equation. If we take a column of an aqueous solution of component i of unit cross-section in the air and suppose that this column is separated across a plane cross-section, two new surfaces are created. In considering this process, we will find at least two types of circumstances. The first is the increasing concentration of solute i in the surface region because of the hydrophobic property of the solute, which is called the adsorption of i at the surface and accompanied by a decrease in surface tension. The second is the decreasing concentration of i in the surface region because of the hydrophilic property of the solute, and the process that is called desorption of i from the surface occurs accompanied by an increasing surface tension. If we take a column of pure water in air and suppose that a component i is introduced at the bottom of the column, then the solute diffuses through the column and also into the surface region. The adsorption of solute i will be observed, regardless of how hydrophobic or hydrophilic the solute is, and the adsorption and desorption processes lead to an equilibrium distribution of i in the surface region. This process has usually been observed by surface tension measurements as a function of concentration. The amount of solute i in the surface region is a positive quantity. However, there is no way to know the amount unless a surface phase or monolayer model is assumed. The measurable quantity in equilibrium is the surface excess of solute and is expressed by the value calculated by the adsorption equation. The positive surface excess at the fluid/fluid interface is called the adsorption, while there are many expressions for negative surface excess of the adsorption: desorption, negative adsorption, surface deficiency, and depletion. This gives us the feeling that we should make clear the meaning of the negative superficial density of a solute. In the first section of this chapter, we consider the adsorption and desorption from binary liquid mixtures in order to describe their surface excess densities and experimental behavior as a function of their composition. Then, upon changing the hydrophobic properties of solutes, we will consider the adsorption from solutions.

4.1   ADSORPTION FROM BINARY LIQUID MIXTURES

The measurements of surface tension of binary liquid mixtures had its beginnings early in the last century. Guggenheim and Adam (1933) discussed the implications of Gibbs' dividing surface and the other choices of the dividing surfaces by means of the surface tension measurements between water–ethanol mixtures and their vapor phases. The positive adsorption of alcohol and the negative adsorption of water were compared for different choices of the dividing surface called conventions. They attempted to clarify the composition in the surface and structure of the surface region. Then the surface tension of binary mixtures has been studied under atmospheric pressure in the presence of air, and the molecular interactions in the surface film have been discussed for many combinations of solvents.

In terms of a simple monolayer model (Defay et al. 1966, p. 168), the surface tension of the perfect solution can be treated as an additive quantity that can be written as

$$\gamma = \gamma_1 x_1 + \gamma_2 x_2 = \gamma_1 - (\gamma_1 - \gamma_2)x_2. \tag{4.1}$$

The monolayer model has been developed in a simple manner to provide a qualitative picture of the experimental data for mixtures. The theory considers a single layer of molecules between two adjacent bulk phases. It is assumed that the layer is an ideal solution and that the chemical potential of the component i in the layer can be represented by

$$\mu_i^m = \mu_i^{m,0}(T,p) + RT \ln x_i^m - \gamma a_i, \tag{4.2}$$

where a_i is a molar surface area. A condition for equilibrium between the layer and bulk liquid mixture is

$$\mu_i^{m,0}(T,p) + RT \ln x_i^m - \gamma a_i = \mu_i^{l,0}(T,p) + RT \ln x_i^l. \tag{4.3}$$

For a pure component i,

$$\mu_i^{m,0}(T,p) - \mu_i^{l,0}(T,p) = \gamma_i a_i. \tag{4.4}$$

Substitution of Equation 4.4 into 4.3 yields the relation

$$x_i^m = x_i^l \exp\left[\frac{(\gamma - \gamma_i)a_i}{RT} \right]. \tag{4.5}$$

For a two-component system, however, the experimental γ versus x_2 curves generally deviate from this straight line. The most common experimental relations are negative deviation from the line and are concave upward. The positive deviation from the curve is also observed for some mixtures (Defay et. al. 1966, Table 12.2; Aveyard 1967).

Various theoretical attempts of giving an explanation of the observed curves for binary mixtures are discussed in many textbooks, but in this section, a discussion of the surface excess quantities obtained from measurements of the surface tension of mixtures of solvents $a1$ and $a2$ will be given. The characteristics of three typical examples of the γ versus x_2 curves are shown in Figure 4.1, in which the lower surface tension component of the mixture is chosen as component a_2. Since, in general, the Gibbs adsorption equation has been used to obtain the surface excess concentrations of the mixtures, let us first consider the variations in Γ_{a1}^{H} and Γ_{a2}^{H}, which can be calculated from Equation 2.93. It is to be noted that the values of Γ_{a1}^{H} represent the surface excess quantity with respect to the two dividing surfaces located so as to make $\Gamma_{air}^{H} = 0$ and $\Gamma_{a2}^{H} = 0$, and that the values of Γ_{a2}^{H} represent those with respect to two dividing surfaces so as to make $\Gamma_{air} = 0$ and $\Gamma_{a1}^{H} = 0$, respectively. In the calculations, mixtures are assumed as the perfect solutions for simplicity. Figures 4.2 through 4.4 illustrate adsorption behaviors observed for three contrasting types of mixtures shown in Figure 4.1. First let us consider the adsorption of isooctane Γ_2^{H} from benzene–isooctane mixtures (Evans Jr. and Clever 1961) whose γ

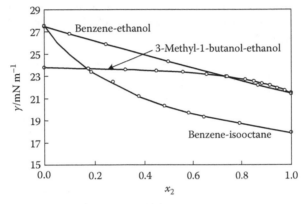

FIGURE 4.1 Three characteristic examples of binary liquid mixtures at 25°C: benzene–ethanol (Myers and Clever 1974), benzene–isooctane (Evans Jr. and Clever 1961), and 3-methyl-1-butanol-ethanol (Alvarez et al. 2011).

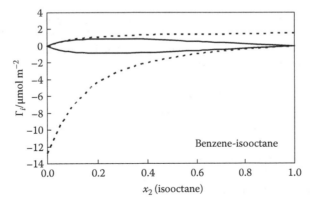

FIGURE 4.2 Comparison of the surface excess concentration–composition curves for Γ_i^{M} and Γ_i^{H} of (1) benzene –(2) isooctane mixtures at 25°C. Upper dotted represents Γ_i^{H} of isooctance; lower one represents that of benzene. Upper and lower solid lines represents Γ_i^{H} of the isooctance and benzene-respectively.

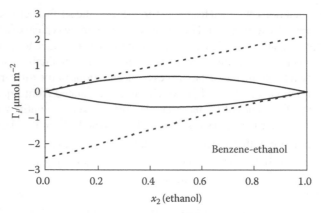

FIGURE 4.3 Comparison of the surface excess concentration–composition curves for Γ_i^M and Γ_i^H of (1) benzene–(2) ethanol mixtures at 25°C. Upper dotted represents Γ_i^H of ethanol; lower one represents that of benzene. Upper and lower solid lines represent Γ_i^M of the ethanol and benzene, respectively.

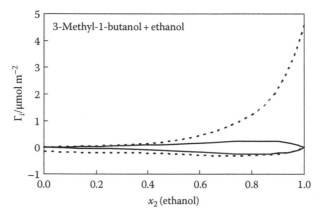

FIGURE 4.4 Comparison of the surface excess concentration–composition curves for Γ_i^M and Γ_i^H of (1) 3-methyl-1-butanol–(2) ethanol mixtures at 25°C. Upper dotted represents Γ_i^H of ethanol; lower one represents that of 3-methyl-1-butanol. Upper and lower solid lines represent Γ_i^M of the ethanol and 3-methyl-1-butanol, respectively.

versus x_{a2} curve is concave upward (Figure 4.2). Since the surface tension of isooctane is lower than that of benzene, isooctane is preferentially adsorbed in the surface region, and the Γ_{a2}^H versus x_{a2} curve is similar in shape to the typical Langmuir adsorption isotherms. The negative Γ_{a1}^H versus x_{a2} curve for benzene in the figure is also similar in shape, but the variation of the curve is much more pronounced than that observed for positive Γ_{a2}^H. Second, in Figure 4.3, Γ_{a1}^H and Γ_{a2}^H of benzene–ethanol mixtures are plotted against the concentration of the ethanol x_{a2} (Myers and Clever 1974). Since the γ versus x_{a2} curve of this mixture is almost a straight line over all compositions, the symmetrical behavior of the adsorption will be attributed to the ideal behavior upon mixing. Figure 4.4 shows the adsorption behavior of the mixture of 3-methyl-1-butanol and the ethanol system (Álvarez et al. 2011) whose γ versus x_{a2} curve is concave to the x_{a2} axis.

It is obvious that the figure is quite similar to that obtained by rotating Figure 4.2 around the origin. Variation in the concentration has a less pronounced effect upon the adsorption of ethanol when $x_2 < 0.5$, but it increases remarkably at a composition higher than 0.5. When we recall that the surface excess concentrations considered earlier are those with respect to the dividing surfaces of solvents, the surface excess concentrations over the whole composition may not be the best attempt for treating the data. It turns out that it is more advantageous to employ dividing surfaces more symmetrical with respect to the two solvents $a1$ and $a2$ (Guggenheim and Adam 1933, Eq. 23).

Next, let us consider the surface excess concentration with respect to two dividing surfaces located so as to make $\Gamma_{air} = 0$ and $\Gamma_{a1} + \Gamma_{a2} = 0$, which was considered in Section 2.9. In Figures 4.2 through 4.4, the calculated Γ_{a1}^M and Γ_{a2}^M are plotted as a function of x_{a2}, where activity coefficients are taken as equal to unity over the whole concentration range. For the choice of two dividing surfaces placed at $V^\sigma = 0$ and $\Gamma_1 + \Gamma_2 = 0$, Guggenheim and Adam (1933) used subscript (N), but the superscript M is used to indicate the convention for the dividing surfaces we used. Since a solvent whose surface tension is higher than the other is always designated as component 1, Γ_{a2}^M is a positive quantity and equal to $-\Gamma_{a1}^M$. When the surface tension is expressed by Equation 4.1, Γ_{a2}^M can be expressed by a symmetrical quadratic curve. The Γ_{a2}^M of the benzene–ethanol mixture in fact shows the quadratic curve (Figure 4.2). The curve skews to the left for the mixtures whose γ versus x_2 curve is concave upward, whereas for the mixtures with downwardly concaved γ versus x_{a2} curve, the curves skew to the right. Further, the experimental data demonstrate that there is a definite correlation between the maximum value of Γ_{a2}^M versus x_{a2} curve and $\Delta\gamma$ between two solvents irrespective of the skewness of the curve.

It is important to note that both Γ_{a2}^H and Γ_{a2}^M are surface excess concentrations of the component $a2$, and the differences in these concentrations are accounted for by the location of the dividing surfaces as defined earlier. The definition of the surface excess concentration relative to the dividing surfaces also accounts for the negative values of Γ_{a1}^H and Γ_{a1}^M. The negative values indicate that the concentration of component 1 is smaller than those in the bulk solution, which is called the surface deficiency of a solute by Langmuir (1917). Further, it should be noted that Γ_{a2}^H and Γ_{a2}^M are nearly equal, and Γ_{a2}^M is equal to $-\Gamma_{a1}^M$ in a dilute solution.

4.2 ADSORPTION FROM SOLUTIONS

Surface tension of binary mixtures is an interesting system theoretically because the variation of the surface tension can be expressed by the molecular interactions in the mixtures of simple molecules. Surface tension of amphiphile solutions shows a number of properties that pure solvents and binary mixtures do not have. One of the properties that belong to the solutions is the formation of a distinctive adsorbed film. Although the state of the film has been studied in detail for insoluble monolayers spread on a water surface, we will find it more convenient to discuss it for the adsorbed film. It is now well established that there exist three distinctive states of the film in the insoluble surface films, that is, gaseous films, expanded films, and condensed films. There is no difference in the properties of the film between the spread monolayer and the adsorbed film, but the thermodynamics for the adsorbed film is well developed by Gibbs.

In this chapter, we will limit our consideration to a three-component two-phase system in which the α phase is the solution of solute i dissolved in a solvent a and the second phase β consists of pure solvent b. When describing the changes in surface tension with the adsorption of the solution, it is convenient to use temperature, pressure, and the chemical potential of the solute. Then,

$$d\gamma = -\Delta s dT + \Delta v dp - \Gamma_i^H \left(\frac{\partial \mu_i}{\partial x_i} \right)_{T,p} dx_i. \tag{4.6}$$

For an ideal dilute solution, the change in the chemical potential of the solute with concentration at constant temperature and pressure is

$$d\mu_i = RTd \ln x_i. \tag{4.7}$$

The surface tension for the ideal dilute solution is given by

$$d\gamma = -\Delta s dT + \Delta v dp - \frac{RT\Gamma_i^H}{x_i} dx_i. \tag{4.8}$$

4.2.1 Adsorption of Octadecanol at Oil/Water Interface

A large number of measurements have been made for an ideal dilute solution of surface active substances, a typical example of which is shown in Figure 4.5 (Matubayasi et al. 1992). In this figure, the surface tensions of the system of water and a benzene solution of octadecanol are plotted. Since the solution is dilute, the measurements of surface tension lowering of this solution at $x = 1 \times 10^3$ show only 0.79 mN m^{-1}, and all plots lie on a linear regression line. Traube (1891) found this rule that the ratio $(\gamma_0 - \gamma)/x$ is independent of the concentration when a solution is dilute. However, the solid line in the figure represents the empirical equation

$$\gamma = C - B \ln(1 + Ax) \tag{4.9}$$

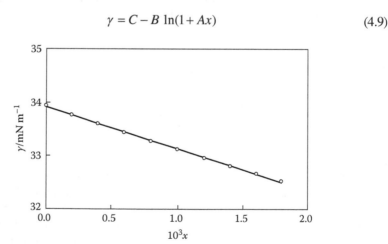

FIGURE 4.5 An example of the simplest surface tension–concentration curve observed for the adsorption of octadecanol at benzene/water interface at 25°C.

in which A, B, and C are constants. This type of equation was found by von Szyszkowski (1908) and used to fit the variation of the surface tension of dilute solutions. He suggested that B is the same for all the fatty acids studied and constant A stands for any one substance. The constant C is the surface tension of a pure solvent. In the discussion to follow Traube's rule and Szyszkowski's equation, Langmuir (1917) has suggested a physical significance of the adsorption. Differentiation of Equation 4.9 and combination with the Gibbs adsorption equation (1.69) yield

$$\Gamma_i^G = \frac{AB}{RT} \frac{x}{(1+Ax)}. \tag{4.10}$$

The right side of this equation has the same form as the Langmuir adsorption equation derived for the adsorption of gases on the solid surface. The derivation is based on the assumption that the enthalpy of desorption is independent of the amount of the adsorbed molecules. For very dilute solutions, Equations 4.9 and 4.10 suggest that γ and Γ_i^G are linear functions of x. Since Γ_i^G has the dimension of number of moles per unit area, the reciprocal of Γ_i^G has been considered as the area occupied by unit mole of the adsorbed molecules A_m. Then, from this relation, we find

$$(\gamma_0 - \gamma)A_m = \pi A_m = RT. \tag{4.11}$$

Since this equation is exactly analogous to the ideal gas law, the adsorbed film in equilibrium with the dilute solution obeys the equation of state of a two-dimensional ideal gas and is commonly called the gaseous adsorbed film. It is to be noted that this two-dimensional ideal gas discarded the term of the solvent when term A_m is introduced and that this A_m is obtained in terms of the surface excess concentration of i rather than the total number of moles of i in the surface region. The values of Γ_i^G and A_m derived from the linear γ–x relation are 0.570 μmol m^{-2} and 2.91 nm^2 molecule^{-1} at $x_i = 1.8 \times 10^{-3}$, respectively.

According to the quasithermodynamic treatment of the surface, the Helmholtz free energy change associated with the adsorption Δf is represented by

$$\Delta f = \Gamma_i^H \left(\left\langle \overline{f}_i^H \right\rangle - \overline{f}_i^\alpha \right) + \Gamma_a^I \left(\left\langle \overline{f}_a^I \right\rangle - \overline{f}_a^\alpha \right) + \Gamma_b^H \left(\left\langle \overline{f}_b^I \right\rangle - \overline{f}_b^\beta \right), \tag{4.12}$$

where Γ_a^I and Γ_b^I are the number of moles of solvents a and b inherent in the interface per unit area. $\left(\left\langle \overline{f}_i^H \right\rangle - \overline{f}_i^\alpha \right)$, $\left(\left\langle \overline{f}_a^I \right\rangle - \overline{f}_a^\alpha \right)$, and $\left(\left\langle \overline{f}_b^I \right\rangle - \overline{f}_b^\beta \right)$ are the partial molar Helmholtz free energy changes of adsorption of components i, a, and b, respectively. In order to examine numerically what part of Δf should be attributed to the solute, we define the apparent molar Helmholtz free energy change Δf^Φ as

$$\Delta f = \Gamma_i^H \Delta f^\Phi + \Delta f_0, \tag{4.13}$$

where the subscript 0 represents the quantity at $\Gamma_i^H = 0$. Since the surface tension equals the change in work content Δf under atmospheric pressure,

$$\Delta f^{\Phi} = \frac{\gamma - \gamma_0}{\Gamma_i^H}. \tag{4.14}$$

At infinite dilution, Δf^{Φ} will approach the partial molar Helmholtz free energy change $\left(\left\langle \overline{f}_i^H \right\rangle - \overline{f}_i^{\alpha}\right)$. In practice, the plots of Δf^{Φ} for the adsorption of octadecanol is particularly interesting because the octadecanol will behave as an ideal solute both in the surface and in the bulk solution. Since the surface tension of octadecanol solution is a linear function of x,

$$\gamma - \gamma_0 = x\frac{d\gamma}{dx}. \tag{4.15}$$

At the temperature of 298.15 K, Equation 4.14 for the gaseous film is

$$\Delta f^{\Phi} = -RT = -2.48\,\text{kJ}\,\text{mol}^{-1}. \tag{4.16}$$

This value is plotted as a dotted line in Figure 4.6. This empirical observation that Δf^{Φ} is independent of the concentration and superficial density of the octadecanol suggests that the adsorbed film is the gaseous film in which the molecules move independently on the surface. In addition to this, Equation 4.16 suggests that the value of $\left(\left\langle \overline{f}_i^H \right\rangle - \overline{f}_i^{\beta}\right)$ is independent of the size and weight of the solute molecules in the gaseous state films whose γ–x relation is straight. The open circles plotted in this figure are Δf^{Φ} values obtained by using γ and Γ_i^H values, which are calculated from Szyszkowski's equation. Equations 4.15 and 4.16 show two facts that the surface tension can be fitted by both the linear equation and Szyszkowski's equation equally well, and that the magnitude of Δf^{Φ} plotted by open circles are almost equal to

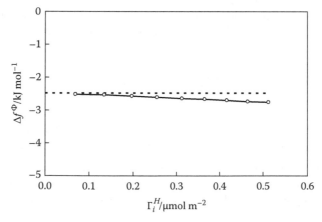

FIGURE 4.6   Variation of apparent molar Helmholtz free energy change Δf^{Φ} for the adsorption of octadecanol at benzene/water interface at 25°C.

the corresponding values obtained by the linear regression line. However, nonlinear regression curve gives rise to the changes in Δf^Φ for the gaseous state film.

A deviation from the linear relation between surface tension and concentration can be observed when benzene is replaced by carbon tetrachloride. The surface tension of CCl_4 solutions of octadecanol at 25°C is plotted against concentration in Figure 4.7, in which a limiting tangent to the γ–x at low concentration is shown as a dotted line to indicate the concentration range where a linear relation holds. From 0 to about 0.5 mol kg^{-1}, the plots fall on a straight line, but thereafter the surface tension deviates upward from the straight line. The solid line connecting the experimental plots seems as if it will be made up of two distinct lines, but all of the plots are well fitted by a single Szyszkowski's equation. At a concentration of 1.67×10^{-3}, the surface tension declines to 4.17 mN m^{-1}, and the adsorption of octadecanol approaches 1.28 μmol m^{-2}. It is to be noted that the linear γ–x relation is important for the gaseous film even though Szyszkowski's equation can be applied to the surface tension data for a wide concentration range.

Figure 4.8 shows the variations with temperature of the γ–x curves of CCl_4 solutions of octadecanol, in which the solid lines represent values calculated by Szyszkowski's equation (Matubayasi et al. 1989). Although the quantitative analysis of the curves is not shown, two typical behaviors of the curves are presented. First, as the temperature is increased, the slope becomes smaller and smaller. Then, at a concentration of 1.67×10^{-3}, the adsorption of 0.89 μmol m^{-2} at the temperature of 35°C is much smaller than that of 25°C. Second, the concentration ranges where the γ–x plots are linear become wider and wider when measurements are made at higher temperatures. The linear part of the γ–x curve at 35°C increased to approximately twofold of that observed at 25°C. It is apparent that, at higher concentrations, the deviation from the linear γ–x curves is due to the presence of molecular interactions between solute molecules in the adsorbed film. Over the limited concentration range in which the ideal gaseous film is observed, the adsorption increases linearly as bulk

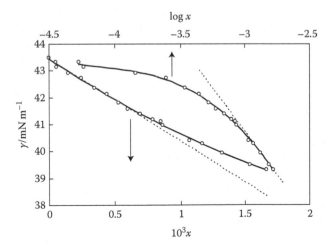

FIGURE 4.7 An example of the deviation from a linear surface tension–concentration curve observed for the adsorption of octadecanol at carbon tetrachloride/water interface.

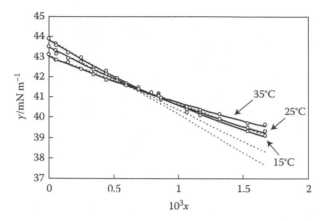

FIGURE 4.8 Temperature dependence of surface tension–concentration curve for the adsorption of octadecanol at carbon tetrachloride/water interface.

concentration increases and thereafter deviates downward from the line according to the familiar behavior of Langmuir's adsorption isotherm.

In Figure 4.7, the surface tension data are also plotted against the natural logarithmic of the mole fraction of octadecanol, because most of the surface tension data of solutions have been presented using semilog plots. In general, a straight line is drawn at the concentration just below near the micelle formation so that the magnitude of the saturated adsorption of the surfactant may be graphically obtained. In spite of this usefulness in determining the slope graphically, however, it is disadvantageous to express the concentration in terms of the logarithmic scale. For example, if we apply the tangent line of γ–log x curve where the saturation of the adsorption of the expanded film is not attained as shown in the figure, the calculated plateau will be a mean value at that concentration range. The γ–log x curves are basically concave to the horizontal axis, but the change in slope becomes unclear by shrinking the x axis at the higher concentration region and by expanding the x axis at the lower concentration region.

The adsorption of octadecanol is even more pronounced at hexane/water interface (Matubayasi et al. 1978). The γ–x curves at several pressures are shown in Figure 4.9. At the atmospheric pressure, the surface tension decreases monotonously with increasing concentration in the concentration range from 0 to 2×10^{-3}. However, the concentration range where the γ–x curve is linear occurs in a very narrow concentration range. An important feature of this figure is a sharp corner on the γ–x curve observed at an elevated pressure. The surface tension first decreases monotonously with concentration, passes through a sharp corner, and thereafter decreases abruptly with concentration. At the concentration range in which there is a steep drop in surface tension, the adsorption of octadecanol reaches a maximum. The values of nearly 8 μmol m⁻² for the monomolecular film show that the area occupied by one molecule is 0.2 nm². This is possible only when the film is exactly composed of octadecanol molecules closely packed and arranged perpendicular to the surface. Evidently, the large amount of the adsorption and the strong lateral interaction may cause abrupt transition between expanded and condensed states of the film. At this transition point, there are significant

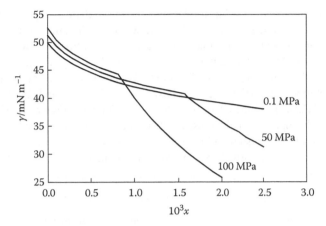

FIGURE 4.9 Pressure dependence of surface tension–concentration curve for the adsorption of octadecanol at hexane/water interface at 30°C.

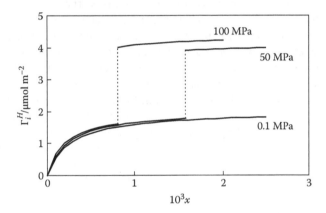

FIGURE 4.10 Pressure dependence of the adsorption of octadecanol at hexane/water interface at 30°C.

discontinuous changes in the adsorption as shown in Figure 4.10. It is clear from this figure and from the plots in Figures 4.6 and 4.7 that there is no experimental observation that suggests the existence of the gaseous and expanded state of the film. Since the change in the surface tension of the gaseous film is limited to about 2 or 3 mN m^{-1}, there is a possibility of insufficient significant figures of the measurements. However, noncritical gaseous expanded transition is likely to occur sequentially above the critical temperature. We could expect to observe the gaseous/expanded transition if we carefully learn about molecules with strong intermolecular forces.

4.2.2 ADSORPTION OF ZWITTERIONIC AMINO ACIDS AT AIR/WATER INTERFACE

As is clear from the γ–x relations in the previous sections, the shape of the curve would be expected to depend upon the van der Waals interactions between solutes–solutes, solvents–solvents, and solutes–solvents in the adsorbed film. In the adsorbed film where solutes–solutes interaction > solutes–solvents interaction, there is apparently a repulsive force between solutes and solvents, and the force would result in the transition of the film. Amino acids are typical amphipathic molecules, and an attractive force between their dipole–dipole would be expected in the adsorbed film (Matubayasi et al. 2002). The surface tension of the aqueous solution of 2-amino-3-methyl-butanoic acid (Val), 2-amino-3-methyl-pentanoic acid (Ile), 2-amino-hexanoic acid (Nle), and 2-amino-4-methyl pentanoic acid (Leu) as a function of concentration is shown in Figure 4.11 (Matubayashi et al. 2007). In this figure, molality is used instead of the mole fraction, and the 0.1 mol kg^{-1} solution is equal to the aqueous solution of $x = 1.8 \times 10^{-3}$. The surface tension of these amino acids decreases linearly from the origin in a manner similar to that observed for octadecanol. The linear γ–m plots of these amino acids suggest that $\Delta f^{\Phi} = -2.48$ kJ mol^{-1} and is equal to the value $\left(\left\langle \overline{f}_i^{H} \right\rangle - \overline{f}_i^{\alpha} \right)$ at the concentration range in which γ–m plots are linear. This result may suggest that the work content associated with the transfer of the amino acids from the bulk water into the surface region is the same as that of the transfer of octadecanol from organic solvents into the surface region. The explanation will not appeal to us who intuitively expect the distinctive properties of solutes and solvents. However, the result is similar to the statement that the ideal gas law $\pi A_m = RT$ will be true in a gaseous film formed at either an air/water interface or an oil/water interface.

The surface tension curves for Leu and its isomers in the figure are particularly interesting as they refer to the transition from gaseous to expanded states of the film. The curve consists of two straight lines connected at a sharp corner, one for the gaseous state and one for the expanded state of the film. In the figure, a tangent line of the gaseous state film is shown as the dotted line. The γ–m plots of the expanded state film deviate downward from the tangent line of the gaseous film and decrease as the concentration increases. This variation is quite contrary to that observed for the γ–x curve shown in Figure 4.7, which deviates upward from the tangent line. The appearances of the transition point where the curve cannot be differentiable even though the curve is

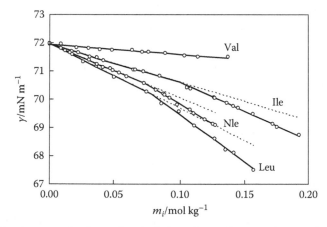

FIGURE 4.11 An example of negative deviations from a linear surface tension–concentration curves observed for valine and leucine isomers at 25°C: 2-amino-3-methyl-butanoic acid (Val), 2-amino-3-methyl-pentanoic acid (Ile), 2-amino-hexanoic acid (Nle), and 2-amino-4-methyl pentanoic acid (Leu).

continuous at the point have a special significance. An abrupt change in the adsorption will occur at the point. This will be due to the fact that the dipole–dipole interactions between amino carboxyl groups are significant in the adsorbed film. This will be supported by the fact that a gaseous/expanded transition cannot be observed for the adsorbed film of sodium leucinate. However, the van der Waals attraction between hydrophobic groups is also important to account for the transition, since the transition point is distinctive for isomers. Especially the transition does not take place in an adsorbed film of tert-leucine (2-amino-3-dimethyl butanoic acid). It is less common to observe the critical gas/expanded transition in the adsorbed film for two reasons. First, in general, the significant figures of the experimental surface tension value are insufficient to ensure the change in the slope of the surface tension–concentration relation. Indeed, the concentration of the transition point depends upon the number of experimental points used to evaluate the intersection of two curves. Second, the attractive force between hydrocarbon chains at the air/water interface probably is insufficient for a convincing account of the transition.

Interaction in the adsorbed film is not the only quantity used for characterizing the adsorption at the surface. The solute and solvent interactions should also be used to account for the adsorption. The solvent effect on the adsorption of octadecanol is a typical example of the close relationship between solvent and solute molecules, but amino acids with a small hydrophobic groups can be used to illustrate a more convincing comparison between the γ–m relations and the solute–solvent interactions. In Figure 4.12, the γ–m relations for the aqueous solution of aminoacetic acid (Gly) and 2-amino-propionic acid (Ala) are compared with those of Val and Leu. It is shown that the surface tensions of the aqueous solutions of these amino acids are sensitive to the change in the length of the hydrocarbon chain of α-amino acid. The γ–m plots of Gly and Ala are slightly concave to the concentration axis, but they practically are straight lines. The $d\gamma/dm$ of these solutions is characteristic for each amino acid and varies with the chain length. They have the same hydrophilic group, the amino-carboxyl group, and their hydrophilic nature becomes significant with decreasing chain length. The adsorption of solute molecules with positive $d\gamma/dm$ gives negative Γ_i^H and is generally called negative adsorption.

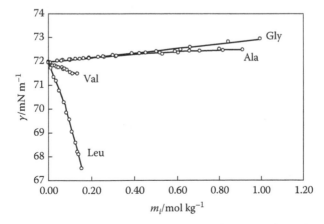

FIGURE 4.12 An example of surface tension–concentration curves for amino acid of which the surface tension increases with increasing concentration.

4.2.3 ADSORPTION OF GLUCOSE, FRUCTOSE, AND SUCROSE AT AIR/WATER INTERFACE

Jones and Ray (1937) and Long and Nutting (1942) measured the γ–m relations of sucrose solutions to compare them with those of inorganic salt solutions. Their measurements are limited to a very low concentration region from 0 to 0.01 M, but their measurements clearly show a hydrophilic nature of sucrose in the aqueous solutions. A linear γ–concentration relation has been presented. In Section A.7, the surface tension of the aqueous solutions of glucose, fructose, and sucrose is shown in Tables A.42 through A.44. Even in this moderately concentrated solution, the surface tension increases linearly as concentration increases as shown in Figure 4.13 (Matubayasi and Nishiyama 2006). The distinctive $d\gamma/dm$ values of glucose, fructose, and sucrose obtained by linear regression line are 1.09, 0.96, and 0.92 mN m^{-1} kg mol^{-1}, respectively. The aqueous solutions of glucose and fructose are a mixture of some isomers of cyclic forms and open-chain forms, but their γ–m plots are strictly straight. Since the $d\gamma/dm$ values of these sugars are similar, the graphs are indistinguishable if their γ–m plots are compared on the figure. These values are very close to the value of 1.04 mN m^{-1} for sodium iodide, but $\left[\left(\Gamma_i^H/m\right)\big/\mathrm{mg}\ \mathrm{m}^{-2}\right]$ values of NaI and sucrose are -0.211 and -0.371, respectively.

4.2.4 ADSORPTION OF SODIUM CHLORIDE AT AIR/WATER INTERFACE

Sodium chloride is a typical example of surface inactive substances, which is often referred to in the textbook (Matubayasi et al. 1999). In Figure 4.14, the γ–m plot of the aqueous solution of sodium chloride is compared with those of glucose and glycine solutions. The γ–m plot of sucrose almost coincides with that of glycine. For 1:1-type electrolyte i, the change in surface tension is given as a function of T, p, and m_i:

$$d\gamma = -\Delta s dT + \Delta v dp - \frac{2RT\Gamma_i^H}{m_i}\left[1+\left(\frac{\partial \ln f_\pm}{\partial \ln m_i}\right)_{T,p}\right]dm_i. \qquad (4.17)$$

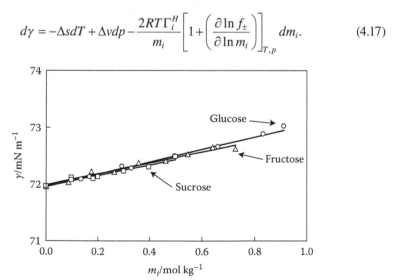

FIGURE 4.13 Surface tension–concentration curves of the aqueous solution of glucose, fructose, and sucrose at 25°C.

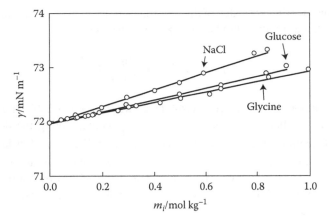

FIGURE 4.14 Comparison of the increments of surface tension observed for the aqueous solution on sodium chloride, glucose, and glycine.

The Γ_i^H values calculated by this equation are plotted against concentration together with those of sucrose and glycine in Figure 4.15. At a lower concentration of 0.3, these three Γ_i^H–m plots almost coincide with each other, and thereafter the plots for glycine deviate downward from the other two curves. The adsorption of sucrose is calculated provided that the solution is an ideal dilute solution. The topmost curve for NaCl is almost linear because the contribution of the activity coefficient is only 3%. When we refer to the order of the slope of the γ–m curves, it seems intuitive that the negative value of Γ_i^H does not correspond to the order of the magnitude of the slopes. This complication arises from the tradition that the concentration of the salt is taken as one of the variables instead of the concentration of ions in the solution. The Γ_i^H calculated by Equation 4.17 is Γ_{NaCl}^H instead of $\left(\Gamma_{Na^+}^H + \Gamma_{Cl^-}^H\right)$.

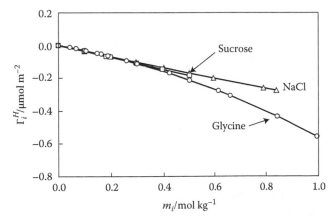

FIGURE 4.15 Comparison of the negative surface excess concentration of sodium chloride, sucrose, and glycine.

Let us consider the Δf^{Φ} value of this salt solution. By means of Equations 4.14 and 4.17 and the linear relationship between γ and m, we have

$$\Delta f^{\Phi} = -2RT\left(1 + \frac{\partial \ln f_{\pm}}{\partial \ln m}\right). \tag{4.18}$$

At infinite dilution, Δf^{Φ} of a 1:1 electrolyte shows the value $-2RT$. This result may formally indicate that the change in $\left(\left\langle \overline{f}_i^{H} \right\rangle - f_i^{\alpha}\right)$ accompanying the desorption of ions is -2.48 kJ mol^{-1} irrespective of the concentration. When we consider Equation 4.13 for the adsorption of surface active substances, we have assumed intuitively that Γ_i^I can be replaced by Γ_i^H. However, we should be aware of the distinction in meaning that is considered in Chapter 2. To avoid possible confusion, it is desirable to rewrite Equation 4.13 as

$$\Delta f = \Gamma_i^I \Delta f^{\Phi} + \Delta f_0 \tag{4.19}$$

and Equation 4.14 as

$$\Delta f^{\Phi} = \frac{\gamma - \gamma_0}{\Gamma_i^I}. \tag{4.20}$$

Δf^{Φ} will approach closely to the partial molar Helmholtz free energy change $\left(\left\langle \overline{f}_i^{I} \right\rangle - \overline{f}_i^{\alpha}\right)$ at infinite dilution. That is, in a sufficiently dilute solution,

$$\gamma - \gamma_0 = \Gamma_i^I\left(\left\langle \overline{f}_i^{I} \right\rangle - \overline{f}_i^{\alpha}\right) = \Gamma_i^H\left(\left\langle \overline{f}_i^{H} \right\rangle - \overline{f}_i^{\alpha}\right). \tag{4.21}$$

Since Γ_i^I is a positive quantity, $\left(\left\langle \overline{f}_i^{I} \right\rangle - \overline{f}_i^{\alpha}\right)$ is positive for the salt solution whose surface tension increases with increasing concentration. In a dilute solution where $\left(\left\langle \overline{f}_i^{I} \right\rangle - \overline{f}_i^{\alpha}\right)$ is kept constant, the concentration of the ions should increase as surface tension increases.

4.2.5 Adsorption of Bile Acids at Air/Water Interface

Most surfactants are classified in terms of their hydrophilic groups, because their properties in an aqueous solution primarily depend on whether they are part of the anionic, cationic, or nonionic group. An aqueous solution of ionic surfactants is an electrolyte solution, which will be the same as that for the inorganic salt solutions. Then, the change in the surface tension of a solution will be relatively large when we plot γ against m because adsorption of anions and cations is the same. The γ–m plots of sodium taurocholate (NaTC) and sodium taurodeoxycholate (NaTDC) solutions at 25°C are illustrated in Figure 4.16, and the corresponding limiting tangents are shown as dotted lines (Matubayasi et al. 1996, 1997). The precise determination of the concentration range of the gaseous film is difficult, but the linear lowering of surface tension is clearly shown in the figure. The departure of the plots from the tangent line upward in the more concentrated range suggests that the transition between

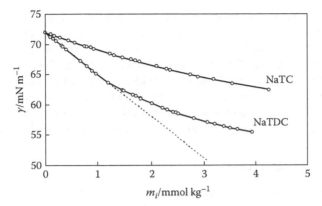

FIGURE 4.16 The surface tension–concentration curves of sodium taurocholate and sodium taurodeoxycholate solutions at 25°C.

gaseous and expanded films occurred gradually, not critically. A gaseous/expanded transition can be more clearly observed for the adsorbed film of dodecylammonium chloride (DAC) at the air/water interface (Motomura et al. 1981). A break point is represented as the intersection of two γ–m lines drawn through the gaseous and expanded states of the films respectively. They also measured the thermodynamic quantities of adsorption of the same surfactant at the hexane/water interface and found no break point on γ–m plots (Aratono et al. 1980). Comparison of these results shows that the structure and the lateral interactions between hydrophobic groups in the adsorbed films are significant along with those of the hydrophilic group. Evidently, the absence of a critical transition for the adsorbed films of NaTC and NaTDC is caused by the less significant lateral interactions in the adsorbed film.

From considerations that the concentrations of the bulk solutions in equilibrium with these adsorbed films are a range of ideal dilute solutions and that the adsorption is derived from the relations between γ and μ_i in the bulk, it should be noted that the shape of the adsorption isotherm is a distinctive property for each surfactant in the adsorbed film. In Figure 4.17, we have plotted Γ_i^H of NaTC against bulk concentration

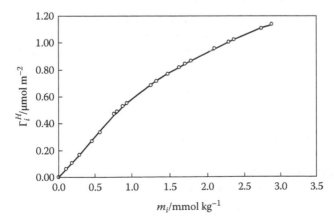

FIGURE 4.17 Adsorption isotherm of sodium taurocholate at 25°C.

and found that NaTC accounts to be a typical adsorption isotherm. The Γ_i^H value first increases linearly with concentration, thereafter the slope gradually decreases as adsorption increases, and finally it becomes flat showing a limited amount of adsorption for an expanded-state film. This saturated adsorption of NaTC is 1.22 µmol m^{-2} while that of DAC is almost 4.5 µmol m^{-2}. Evidently, the adsorption of the bile acid is not remarkable for ionic surfactants. However, the linear γ–m relation of the gaseous film suggests that Δf^Φ is identical with that of DAC. The work content associated with the adsorption does not explain the distinctive features of each surfactant.

4.3 SURFACE ACTIVITY

Traube (1891) has shown some general conclusions based on the extensive measurements of the surface tension of solutions of many organic compounds. First, for many dilute solutions, surface tension decreases linearly as concentration increases. Based on this empirical illustration, Langmuir has considered the properties of gaseous adsorbed film of which may be the two-dimensional ideal gas equation. For the adsorbed film formed on the dilute solution, there is no discrepancy in properties for any one solute. Second, the $-d\gamma/dm$ for dilute solutions is proportional to the chain length of the hydrocarbon chain. This empirical illustration of the aqueous solution of fatty acids is known as Traube's rule. Then the magnitude of $-d\gamma/dm$ is often used as the "surface activity" of a solute, though there is no clear definition of the "surface activity." Upon rearrangement of the adsorption, we have

$$\frac{\Gamma_i^H}{m} = -\frac{1}{RT}\left(\frac{\partial \gamma}{\partial m}\right)_{T,p}. \tag{4.22}$$

The surface activity evaluated at infinite dilution is a distinctive property for any one of the solutes.

Equation 4.19 suggests that a more adequate designation would be a value of Γ_i^H/m that represents the rate of surface density with respect to the bulk concentration.

These considerations raise an important question. Is $\left(\left\langle \bar{f}_i^H \right\rangle - \bar{f}_i^\alpha\right)$ actually a colligative property and why is the individuality of the solute not observed? We know from the experiment that partial molar volume in a dilute solution is specific for a solute. Recalling the definitional equation of equilibrium (2.110), we have Equation 2.111 for Δf as

$$\Delta f = \gamma - p\Delta v.$$

In Equations 2.14 and 2.20, we have assumed that the $p\Delta v$ term can be neglected for the condensed phase. As we have seen in Table 3.3, where the magnitudes of γ, $p\Delta v$, and $T\Delta s$ are shown, the $p\Delta v$ term is negligible compared with γ and $T\Delta s$ for a two-phase two-component system. Further, we suppose that the variation of γ with concentration is limited and insensitive for the variation of species. Thus, $\left(\left\langle \bar{f}_i^H \right\rangle - \bar{f}_i^\alpha\right)$ is evidently insensitive to the variation of species in dilute solutions as long as γ varies linearly with concentration and Γ_i^H/m is a specific quantity for a specific solute.

4.4 ENTROPY CHANGE ASSOCIATED WITH THE ADSORPTION

4.4.1 SURFACE ACTIVE MOLECULES

The entropy change associated with adsorption can be evaluated by

$$\Delta s = -\left(\frac{\partial \gamma}{\partial T}\right)_{p,m}$$

from the experimental slopes of γ–T plots of solutions. Using the plots of octadecanol at a CCl$_4$/water interface, the characteristics of γ–T curves over a limited temperature range (20°C–35°C) are shown in Figure 4.18. For pure and dilute solutions, the slopes $d\gamma/dT$ are negative and the surface tension is a linear function of temperature in this narrow temperature range. As concentration increases, the slope passes through zero and turns positive. Figure 4.19 illustrates the variation of Δs against m. The positive

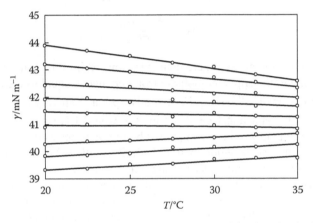

FIGURE 4.18 An example of the linear γ–T plots observed for the adsorbed film of octadecanol at the CCl$_4$/water interface.

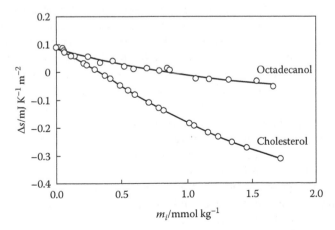

FIGURE 4.19 The entropy of surface formation Δs of octadecanol and cholesterol at the CCl$_4$/water interface.

value of Δs for a pure CCl_4/water interface decreases linearly with the addition of octadecanol in the concentration range where a linear γ–m is observed and thereafter the rate of the decrease is slowed on further increasing concentration. For the adsorption of a surface active solute i at the interface between fluids a and b, a better understanding will be achieved by considering the relation

$$\Delta s = \Gamma_a^I \left(\left\langle \overline{s}_a^I \right\rangle - \overline{s}_a^\alpha \right) + \Gamma_b^I \left(\left\langle \overline{s}_b^I \right\rangle - s_b^\beta \right) + \Gamma_i^H \left(\left\langle \overline{s}_i^H \right\rangle - \overline{s}_i^\alpha \right). \tag{4.23}$$

The first two terms on the right of this equation are the partial molar entropy changes of the adsorption of solvents a and b. The decreases in Δs for dilute solutions appear definitely to be the negative contribution of the third term. The partial molar entropy of octadecanol in the surface region will be less than that in the bulk phase when the entire system is in equilibrium. In order to determine the difference in $\left(\left\langle \overline{s}_i^H \right\rangle - \overline{s}_i^\alpha \right)$ numerically, let us define the apparent molar entropy change Δs^ϕ by the expression

$$\Delta s^\phi = \frac{\Delta s - \Delta s_0}{\Gamma_i^H}, \tag{4.24}$$

where Δs_0 is the value of Δs at zero concentration. At infinite dilution where γ–Γ_i^H relations are linear, the apparent molar entropy change is equal to the change in the partial molar entropy $\left(\left\langle \overline{s}_i^H \right\rangle - \overline{s}_i^\alpha \right)$. For the adsorption from very dilute octadecanol solution at a CCl_4/water interface, Δs^ϕ shows -0.096 kJ K^{-1} mol^{-1}. It seems likely that the negative value will be accompanied by the more restrictive conformation of octadecanol in the surface region than in the bulk solution (Motomura et al. 1978).

A more simple and useful understanding will be obtained by considering the relationship between the magnitude of Δs^ϕ and the structure of the hydrophobic group. In Figure 4.19, plots of Δs against m for the adsorption of cholesterol at CCl_4/water interface are compared with that of octadecanol. At the concentration range where a gaseous adsorbed film is formed, the values of Δs of these two solutes decrease linearly on increasing the concentration. It appears probable that no appreciable lateral interaction between adsorbed molecules may be expected. The Γ_i^H / x_i values of octadecanol and cholesterol are respectively 1.4 and 2.7 μmol m^{-2}, and Δs of cholesterol would descend much more steeply than that of octadecanol. When we compare the chemical structure of sterol and hydrocarbon chain, the specific dependence of Δs^ϕ upon the chemical structure would be expected to be appreciable. However, results derived from the experimental data show that both are of the same magnitude of Δs^ϕ (ca. -0.096 kJ K^{-1} mol^{-1}). The attractive force between OH group and water will be significant, but since both dissolve well in CCl_4, it is questionable whether the hydrophobic groups take very restricted conformation in the surface region so that the distinction between them can be evaluated from the magnitude of Δs^ϕ.

Let us next consider the adsorption of surfactants from an aqueous solution on the air/water interface. In Figure 4.20, the surface tension is plotted against temperature for the aqueous solutions of NaTC between 15°C and 35°C. The surface tension of the air/solution interface can be treated as a linear function of temperature in this range of temperature and concentration. By using the method of least squares, the calculated Δs is plotted against m in Figure 4.21. As we have pointed out in the previous chapter, the slope for a pure air/water interface is larger than that

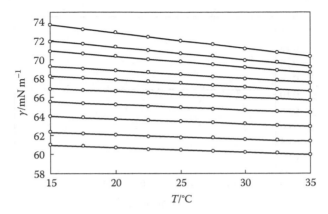

FIGURE 4.20 Linear γ–T plots observed for the aqueous solutions of NaTC.

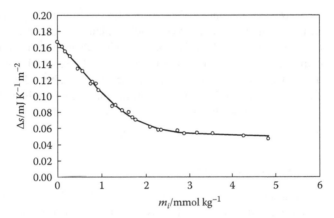

FIGURE 4.21 Entropy change of adsorption of sodium taurocholate at air/water interface.

of an oil/water interface because of the mixing in the surface region (Table 3.3). For dilute solutions, the decrease of Δs is proportional to concentration; thereafter, the decrease ceases as concentration increases and becomes flat where a saturated expanded film is formed. The shape of the Δs–m plot is similar to that of the adsorption at the CCl_4/water interface, but the variation in Δs has been much more pronounced, though it does not show negative values. In comparing the adsorption from water and from oil phases, we are interested in changes in the molar contribution of surfactant $\left(\left\langle \bar{s}_i^H \right\rangle - \bar{s}_i^\alpha \right)$. The Δs of an electrolyte solution will be written in the form

$$\Delta s = \Gamma_a^I \left(\left\langle \bar{s}_a^I \right\rangle - \bar{s}_a^\alpha \right) + \Gamma_b^I \left(\left\langle \bar{s}_b^I \right\rangle - s_b^\beta \right) + \Gamma_+^H \left(\left\langle \bar{s}_+^H \right\rangle - \bar{s}_+^\alpha \right) + \Gamma_-^H \left(\left\langle \bar{s}_-^H \right\rangle - \bar{s}_-^\alpha \right). \quad (4.25)$$

By making use of Equation 2.157, a form similar to Equation 4.23 can be derived. In Figure 4.22, Δs^ϕ are plotted against Γ_i^H. Here the values of Γ_i^H are calculated using three empirical relations: linear, Szyszkowski, and polynomial equations for dilute, intermediate, and more concentrated solutions, respectively. The open circles in the figure are obtained by using the values plotted in Figure 4.20, and the solid line is

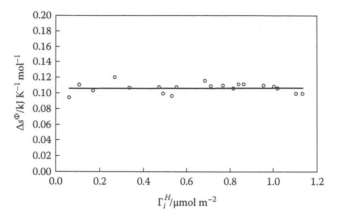

FIGURE 4.22 Apparent molar entropy change of adsorption Δs^ϕ of sodium taurocholate.

obtained by using the best-fit line of the Δs–m plot. Although the plots for dilute solutions are scattered, it appears that Δs^ϕ is independent for adsorption up to about 1 μmol m^{-2}. The value of Δs^ϕ of NaTC associated with adsorption from the aqueous solution to the air/water interface is found to be -0.106 kJ K^{-1} mol^{-1}, while that of octadecanol from CCl$_4$ solution to CCl$_4$/water interface is -0.096 kJ K^{-1} mol^{-1}. The difference in their magnitude, 0.011, may be significant, but it is to be noted that the adsorptions of surfactants both from the oil phase and from the water phase show almost similar magnitude of Δs^ϕ, that is, $\left(\left\langle \overline{s_i}^H \right\rangle - \overline{s_i}^\alpha \right)$.

4.4.2 ZWITTERIONIC AMINO ACIDS

The Δs^ϕ values for amino acids exhibit a curious feature. Leu, 2-amino-4-methyl pentanoic acid, is a surface active compound whose adsorbed film shows a clear gaseous/expanded transition (Figure 4.11). The Δs value of Leu is calculated using γ–T data and plotted against m in Figure 4.23. In comparing this figure with Figures 4.19 and 4.21,

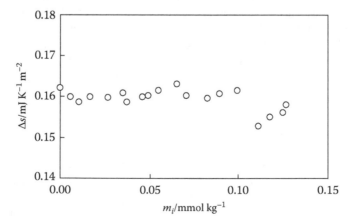

FIGURE 4.23 Entropy of adsorption of leucine.

it can be noted that the plots of Δs against m for Leu would be a horizontal straight line. A possible explanation is that $\left\langle \overline{s}_i^H \right\rangle$ will have a value comparable to $\overline{s}_i^\alpha$ in the aqueous solution, and a possible difference between NaTC and Leu is the size of their hydrophobic group. A conclusion from these observations is that transition between the gaseous and expanded film of amino acid cannot be observed if the electrostatic attractive force is not significant.

4.4.3 SUCROSE

The earlier experimental results suggest that a hydrophobic group of small size would be expected to show no important effect on the magnitude of Δs. This idea of the magnitude of the effect of hydrophilic group will be gained from the inspection of the Δs values of the hydrophilic molecules such as sucrose. The sucrose molecules are hydrophilic enough to increase the surface tension of their aqueous solutions. Table 4.1 shows that the Δs of the aqueous solution of sucrose is practically independent of the bulk concentration and of the negative surface excess density. Thus, recalling Equation 4.23, we may write $\Gamma_i^H \left(\left\langle \overline{s}_i^H \right\rangle - \overline{s}_i^\alpha \right) = 0$. The partial molar entropy of sucrose in the surface region should be the same as that in the bulk solution whatever negative values Γ_i^H takes. When phases α and β are practically immiscible and solute i is soluble only in phase α, we found from Equations 2.114 and 2.117 that

$$\Gamma_i^H \left(\left\langle \overline{s}_i^H \right\rangle - \overline{s}_i^\alpha \right) = \Gamma_i^I \left(\left\langle \overline{s}_i^I \right\rangle - \overline{s}_i^\alpha \right). \tag{4.26}$$

According to the consideration in Section 4.2.4, Γ_i^I should have nonzero positive value, and this relation indicates that $\left\langle \overline{s}_i^I \right\rangle = \left\langle \overline{s}_i^H \right\rangle = \overline{s}_i^\alpha$. It should be noted that hydration of sucrose and glucose in the surface region does not differ from that observed in the bulk solution, although sucrose has no ionic charge in the structure. This consideration may hold for the adsorption of amino acid solutions.

The Δs–m relation observed for sucrose solutions suggests that the first and second terms in Equation 4.23 remain almost constant. In general, we expect some

---

TABLE 4.1
Entropy of Adsorption Δs and Adsorption Γ_i^H of Sucrose and Glucose at 25°C

	Sucrose			Glucose	
m (mol kg^{-1})	Δs (mJ K^{-1} m^{-2})	Γ_i^H (μmol m^{-2})	m (mol kg^{-1})	Δs (mJ K^{-1} m^{-2})	Γ_i^H (μmol m^{-2})
0.0000	0.162	0.000	0.0000	0.162	0.000
0.0979	0.160	−0.036	0.1349	0.161	−0.059
0.1812	0.160	−0.067	0.1679	0.159	−0.074
0.1992	0.163	−0.074	0.2932	0.160	−0.129
0.2989	0.161	−0.111	0.4986	0.161	−0.219
0.3974	0.161	−0.147	0.6601	0.161	−0.291
0.5001	0.161	−0.186	0.8335	0.161	−0.367

variation in partial molar quantities given for the aqueous solutions of moderately concentrated solutions. Even when the third term can be explained by the hydration of sucrose, the constancy of the first and second terms is noteworthy.

4.4.4 SODIUM CHLORIDE

The preceding observations for the adsorption of zwitterions and the surface deficiency of sucrose suggest that the hydration of ions and the negative surface excess density would not accompany the changes in Δs. In general, two outlooks for the surface tension of the aqueous solutions of simple electrolytes are adopted. First, the temperature coefficient of γ is practically independent of concentration, and the departure of the Δs of electrolyte solution from pure water will be small (Lyklema 2000). Second, the temperature effect of $d\gamma/dm$ is smaller than the experimental error of the surface tension (Weissenborn and Pugh 1996). These expectations ensure the consistency of the relation

$$\frac{d}{dm}\left(\frac{d\gamma}{dT}\right) = \frac{d}{dT}\left(\frac{d\gamma}{dm}\right). \tag{4.27}$$

However, we found that the Δs of the aqueous solution of NaCl clearly decreases with increasing concentration, although the negative surface excess density and sufficient hydration are expected (Table 4.2).

In the absence of sufficient experimental data on the electrolyte solutions, it would be unwise to attribute any observation to these outlooks. The experimental works have been focused mainly on determining the most probable magnitude of $d\gamma/dm$, since repulsion of ions from the surface has been of theoretical interest. The values of $d\gamma/dm$ of electrolyte solutions have been reported repeatedly usually with three significant figures, but have different values for different observers. Thus, experimental verifications of the reciprocity of the surface tension with respect to concentration and temperature are not common. As an example of the relation, the $d\gamma/dm$ of NaCl is plotted against temperature in Figure 4.24. It is very clear that $d\gamma/dm$ increases with

TABLE 4.2
Surface Tension, Entropy of Adsorption, and Adsorption of Sodium Chloride at 25°C

m (mol kg^{-1})	γ (mN m^{-1})	Δs (mJ K^{-1} m^{-2})	Γ_i^H (μmol m^{-2})
0	71.97	0.162	0.000
0.0990	72.11	0.162	−0.033
0.1982	72.25	0.160	−0.068
0.2951	72.44	0.156	−0.101
0.4012	72.56	0.158	−0.136
0.4981	72.71	0.157	−0.168
0.5937	72.89	0.158	−0.199
0.7899	73.25	0.154	−0.260
0.8389	73.32	0.152	−0.275
1.0121	73.57	0.154	−0.326

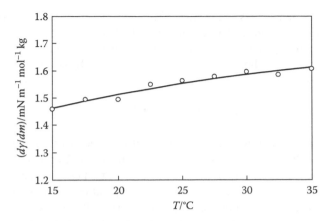

FIGURE 4.24 The $d\gamma/dm$ plot for the aqueous solution of sodium chloride.

temperature, in accord with the data shown in Table 4.2. It will be shown in later chapters that the $d\gamma/dm$ versus temperature relation shown in this figure is typical of the aqueous solution of a single electrolyte without exception. Then, the data shown in Table 4.2 are the typical Δs versus m relations for the aqueous electrolyte solutions that show negative departures from the Δs^0 of pure water.

The evaluated Δs^ϕ using tangential Δs–m relation at infinite dilution shows a positive value about 0.04 kJ K^{-1} mol^{-1}. Since the Δs–m plot is not a straight line, Δs^ϕ decreases further with increasing concentration. The positive values indicate that the $\left\langle \overline{s_i}^H \right\rangle$ of NaCl in the surface region is larger than $\overline{s_i}^\alpha$ in the bulk phase. Unfortunately, the implications of these quantities are not clear to give us unambiguous information about the properties of ions in the surface region. However, Equation 4.26 is helpful to suggest that $\left\langle \overline{s_i}^I \right\rangle$ should be smaller than $\overline{s_i}^\alpha$, since Γ_i^I should be positive. It is hardly to be expected that the negative value of Γ_i^H is sufficient to give a finite extent of the salt-free surface region. This negative Γ_i^H value indicates that the concentration of salt in the surface region is less than that in the bulk phase and that the repulsive forces between ions and the surface is more than sufficient to account for the contribution corresponding to the negative Γ_i^H value but for Γ_i^I. As considered in Section 4.2.4, Γ_i^I should be positive and increases with increasing bulk concentration; thus ions of sodium chloride tend to decrease Δs more and more as the adsorption of ions increases.

4.5 ENERGY, HEAT CONTENT, AND WORK CONTENT ASSOCIATED WITH THE ADSORPTION

When two systems of unequal temperature, pressure, and chemical potential are brought into contact, both systems undergo changes resulting in a state of thermodynamic equilibrium. The transfer of heat may be caused by unequal temperature, the volume change may be caused by unequal pressure, and the transfer of molecules may be caused by unequal chemical potential. The thermodynamic quantities associated with the adsorption Δy are given by

$$\Delta y = \Gamma_a^I \left(\left\langle \overline{y_a}^I \right\rangle - \overline{y_a}^\alpha \right) + \Gamma_b^I \left(\left\langle \overline{y_b}^I \right\rangle - \overline{y_b}^\beta \right) + \Gamma_i^H \left(\left\langle \overline{y_i}^H \right\rangle - \overline{y_i}^\alpha \right). \tag{4.28}$$

At equilibrium, the changes in the Gibbs free energies Δg should be zero, but other thermodynamic quantities differ depending on the temperature, pressure, and chemical potential of the system. In this section, let us consider differences in the internal energy, enthalpy, and Helmholtz work function between the surface region and bulk phase. Δu is given as

$$\Delta u = \Delta f + T\Delta s$$

Here, it is to be noted that Δy is the difference between the thermodynamic quantities in the bulk and those in the surface region in equilibrium. Let us first consider the thermodynamic quantities of cholesterol in equilibrium between the surface region and the CCl_4 solution, since the variations of these quantities of cholesterol are larger than those of other species considered earlier (Figure 4.25). The value of pure CCl_4/ water interface plotted at zero concentration shows that energy, heat content, and work content of surface formation have a positive value. The partial molar quantities of these are larger than those in the bulk phase (Equation 4.28). When cholesterol is introduced into the CCl_4 phase under the fixed temperature and pressure, adsorption of cholesterol occurs until the chemical potential of the cholesterol in the surface region becomes equal to that in the bulk solution. As a result of the adsorption, negative departures of the energy, heat content, and work content from those of the pure interface are observed. When more and more solute is added to the solution, these values become smaller and smaller with increasing concentration. Since Δf^ϕ and Δs^ϕ of the gaseous film are negative and unchanged as described in the previous sections, these descending variations are ascribed to the increasing concentration in the surface region. Then these curves suggest two interesting points. First, the variation of the work content with the adsorption is very small, and the contribution to the lowering of the energy change is less than 10%. There is a significant contribution of the heat content associated with the adsorption. The energetically stable adsorbed film is formed even at the concentration range where the gaseous film is formed. Second,

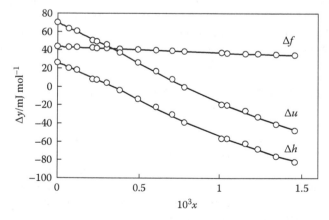

FIGURE 4.25 Thermodynamic quantities of adsorption of cholesterol at the CCl_4/water interface: energy change of adsorption Δu, enthalpy change of adsorption Δh, and Helmholtz free energy change of adsorption Δf, respectively.

the figure shows that there is an energetically stable adsorbed film in equilibrium with the bulk phases, which is accompanied by the negative Δu, even though the surface tension of the solution does not approach zero. The Δu^{Φ} for the adsorption of cholesterol in the gaseous film shows -31 kJ mol^{-1}, while Δf^{Φ} is -2.5 kJ mol^{-1}.

The Δu^{Φ} for the adsorption of NaTC into the gaseous film formed at the air/water interface shows -36.5 kJ mol^{-1}, which is close to the value of the cholesterol, but NaTC never shows negative Δu because of the formation of saturated adsorbed film and the formation of aggregate in the bulk solution on increasing concentration.

There is another type of soluble molecule that can be compiled based on its insignificant surface activity. The enthalpy change associated with the adsorption Δh of sucrose is independent of the concentration; the adsorption or desorption is dependent upon the magnitude of the work content Δf. In order to keep the equilibrium condition $\Delta g = 0$, Δu should be increased with increasing concentration. Because the $d\gamma/dT$ of the solution of leucine is independent of the concentration, the Δh of the solution does not contribute to the adsorption of leucine. Since the positive adsorption of leucine accompanies negative Δf, Δu will be negative. The partial molar energy of leucine in the surface region is always smaller than that in the bulk phase. The relation between these heat and work contents of sodium chloride solution differs from the earlier examples. Figure 4.26 shows the Δy of NaCl solutions as a function of bulk concentration. The work content increases with increasing concentration and contrasts with the descending variation of the heat content. As a result, the energy slightly decreases as concentration in the bulk increases. It is important to note that the work content, heat content, and the energy of adsorption reflect not only the adsorption process but also the effect of chemical structure that the solvent–solute and solute–solute interaction have on the adsorption of the solute. In general, the adsorption at the fluid/fluid interface has been studied as a function of concentration only. Chemical analysis of the concentration of constituent species has been a traditional method in chemistry, but thermodynamic quantities are helpful in understanding the system.

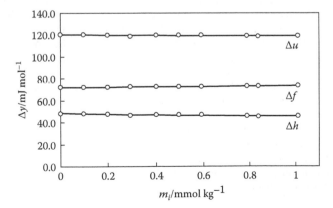

FIGURE 4.26 Thermodynamic quantities of adsorption of sodium chloride at the air/water interface: energy change of adsorption Δu, enthalpy change of adsorption Δh, and Helmholtz free energy change of adsorption Δf, respectively.

REFERENCES

Álvarez, E. A., J. M. Correa, C. E. García-Rosello, and J. M. Navaza. 2011. Surface tensions of three amyl alcohol + ethanol binary mixtures from (293.15 to 323.15) K. *J. Chem. Eng. Data* 56: 4235–4238.

Aratono, M., M. Yamanaka, N. Matubayasi, K. Motomura, and R. Matuura. 1980. Thermodynamic study on the adsorption of dodecylammonium chloride at hexane/water interface. *J. Colloid Interface Sci.* 74: 489–494.

Aveyard, R. 1967. Adsorption from some n-alkane mixtures at the liquid/vapour, liquid/water, and liquid/solid interface. *J. Chem. Soc., Faraday Trans.* 63: 2778–2788.

Defay, R., I. Prigogine, and A. Bellmans. 1966. *Surface Tension and Adsorption*. Longmans, London, U.K.

Evans, Jr., H. B. and H. L. Clever. 1961. Surface tensions of binary mixtures of isooctane with benzene, cyclohexane, and n-dodecane at 30°C. *J. Phys. Chem.* 68: 3433–3435.

Guggenheim, E. A. and N. K. Adam. 1933. The thermodynamics of adsorption at the surface of solutions. *Proc. R. Soc (Lond.) A*139: 218–236.

Jones, G. and W. A. Ray. 1937. The surface tension of solutions of electrolytes as a function of concentration. I. A differential method for measuring relative surface tension. *J. Am. Chem. Soc.* 59: 187–198.

Langmuir, I. 1917. Fundamental properties of solid and liquid II. *J. Am. Chem. Soc.* 39: 1848–1906.

Long, F. A. and G. C. Nutting. 1942. The relative surface tension of potassium chloride solutions by a differential bubble pressure method. *J. Am. Chem. Soc.* 64: 2476–2482.

Lyklema, J. 2000. *Fundamentals of Interface and Colloid Science*, Vol. III. Academic Press, London, U.K.

Matubayasi, N., S. Azumaya, K. Kanaya, and K. Motomura. 1992. Gaseous mixed adsorbed films of octadecanol and cholesterol at the oil/water interface. *Langmuir* 8: 1980–1983.

Matubayasi, N., M. Kanzaki, S. Sugiyama, and A. Matuzawa. 1996. Thermodynamic study of gaseous adsorbed films of sodium taurocholate at the air/water Interface. *Langmuir* 12: 1860–1862.

Matubayasi, N., R. Matsunaga, and K. Motomura. 1989. Interaction of cholesterol and octadecanol in a mixed adsorbed film at carbon tetrachloride water interface: Criticism about the condensing effect of cholesterol. *Langmuir* 5: 1048–1051.

Matubayasi, N., H. Matsuo, K. Yamamoto, S. Yamaguchi, and A. Matuzawa. 1999. Thermodynamic quantities of surface formation of aqueous electrolyte solutions I. Aqueous solutions of NaCl, MgCl$_2$, and LaCl$_3$. *J. Colloid Interface Sci.* 209: 398–402.

Matubayasi, N., S. Matsuyama, and J. Zhang. 2007. Unpublished data.

Matubayasi, N., H. Miyamoto, K. Yano, and T. Tanaka. 2002. Thermodynamic quantities of surface formation of aqueous electrolyte solutions V. Aqueous solution of aliphatic amino acids. *J. Colloid Interface Sci.* 250: 431–437.

Matubayasi, N., K. Motomura, M. Aratono, and R. Matuura. 1978. Thermodynamic study on the adsorption of 1-octadecanol at hexane/water interface. *Bull. Chem. Soc. Jpn.* 51: 2800–2803.

Matubayasi, N. and I. Nishiyama. 2006. Thermodynamic quantities of surface formation of aqueous electrolyte solutions VI. Comparison with typical nonelectrolytes, sucrose and glucose. *J. Colloid Interface Sci.* 298: 910–913.

Matubayasi, N., S. Sugiyama, M. Kanzaki, and A. Matuzawa. 1997. Thermodynamic studies of the adsorbed films and micelles of sodium taurodeoxycholate. *J. Colloid Interface Sci.* 196: 123–127.

Motomura, K., S. Iwanaga, Y. Hayami, S. Uryu, and R. Matuura. 1981. Thermodynamic studies on adsorption at interfaces IV. Dodecylammonium chloride at water/air interface. *J. Colloid Interface Sci.* 80: 32–38.

Motomura, K., N. Matubayasi, M. Aratono, and R. Matuura. 1978. Thermodynamic studies on adsorption at interfaces II. One-surface active component system: Tetradecanol at hexane/water interface. *J. Colloid Interface Sci.* 64: 356–341.

Myers, R. S. and H. L. Clever. 1974. The surface tension and density of some hydrocarbon + alcohol mixtures at 303.15 K. *J. Chem. Thermodyn.* 6: 949–955.

Szyszkowski, B. von. 1908. Experimental studies of the capillary properties of aqueous solutions of fatty acids. *Z. Physik. Chem.* 64: 385–414.

Traube, I. 1891. Ueber die capilaritätsconstanten organischer stoffe in wässerigen lözungen. *Ann. Chem. Liebigs* 265: 27–55.

Weissenborn, P. K. and R. J. Pugh. 1996. Surface tension of aqueous solutions of electrolytes: Relationship with ion hydration, oxygen solubility, and bubble coalescence. *J. Colloid Interface Sci.* 184: 550–563.

5 Surface Tension of Simple Salt Solutions

In this chapter, we consider the $\gamma-m$ relations and their slopes $d\gamma/dm$ of the aqueous solutions of simple salts. From the very beginning of the study of electrolytes, the profiles of the $\gamma-m$ plots and their $d\gamma/dm$ values are of particular interest. Thus far, our attention has been restricted to find and to confirm a specific factor with which the magnitude of the $d\gamma/dm$ is attainable. This chapter reviews experimental inroads into the relationship between hydration forces and the surface tensions of simple electrolytes at 25°C.

5.1 AQUEOUS SOLUTIONS OF 1:1 ELECTROLYTES

5.1.1 ALKALI METAL HALIDES

Alkali halides are typical simple salts and have been studied repeatedly by many researchers. The $\gamma-m$ relations of sodium halides and alkali metal chlorides are plotted in Figures 5.1 and 5.2 (Matubayasi et al. 1999, 2001). These figures illustrate the following three important features:

1. The surface tension increases almost linearly with the addition of salt into the bulk water. The increase indicates that the ions are adsorbed into the surface region proportional to the added amount of salt and interact with the surface water. There is a deficiency of ions in the surface region compared with the bulk solution, but the adsorption occurs so that the isotropy of the chemical potential of each constituent is attained through the system. At constant T and p, Equation 2.14 reduces to

$$\Gamma_i^H = -\frac{m_i}{RT(v_+ + v_-)}\frac{d\gamma}{dm_i}\bigg/\left(1+\frac{d\ln f_\pm}{d\ln m_i}\right). \tag{5.1}$$

The negative surface excess density reaches -0.34 μmol m^{-2} for 1.0 mol kg^{-1} NaCl solutions. This large negative value suggests that we can scarcely expect to find ions in the surface region, since we know from experience that saturated adsorption of surfactants will reach 1–2 μmol m^{-2}. However, the limited amount of ions significantly contributes to the increments of the surface tension as concerned in the previous chapter. The inspection of $\gamma-m$ graphs and the significant digits of the data permit us to calculate the slope of the straight line as an index of the capacity to elevate the surface tension, as an index of the deficiency of ions, and as an index of the hydration of ions.

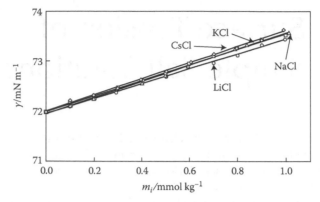

FIGURE 5.1 Comparison of surface tension–concentration curves of alkali metal chlorides at 25°C: circles (LiCl), triangles (NaCl), squares (KCl), and diamonds (CsCl).

2. Effect of cations on the magnitude of $d\gamma/dm$ is primarily a function of the number of charges on each cation, because γ–m relations of lithium, sodium, potassium, and cesium chlorides almost coincide with each other as shown in Figure 5.1. This result is in agreement with the theoretical consideration that the repulsion from the image force is responsible for the increment of the surface tension. The magnitude of $d\gamma/dm$ of cations is in the order CsCl > KCl ≃ NaCl > LiCl, although this is exactly opposite of the order observed by Weissenborn and Pugh (1996). The difference between values of CsCl and LiCl (about 0.2 mN m^{-1} mol^{-1} kg) is small but significant, so it seems possible that ions with smaller hydration radii will also have a larger effect on the surface tension increments even if we consider the magnitude of the experimental error.

3. Effects of anions on the magnitude of $d\gamma/dm$ are clearly different in appearance compared with those of cations. Figure 5.2 illustrates that the magnitude of $d\gamma/dm$ of sodium halides is sensitive to the extent of hydration of anions and is in the order NaF > NaCl > NaBr > NaI. It depends on some physical properties in addition to the effect of the charge number of ions, because $d\gamma/dm$ varies ion by ion.

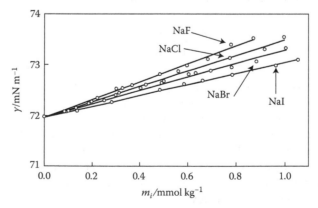

FIGURE 5.2 Comparison of surface tension–concentration curves for sodium fluoride, chloride, bromide, and iodide from top to bottom at 25°C.

TABLE 5.1
[($d\gamma/dm$)/mN m^{-1} mol^{-1} kg]
Values of Alkali Metal
Halides at 25°C

	Li$^+$	Na$^+$	K$^+$	Cs$^+$
F$^-$	—	1.79	—	—
Cl$^-$	1.42	1.55	1.57	1.63
Br$^-$	—	1.23	1.29	—
I$^-$	—	1.04	1.13	—

The intersection of a particular row of cation and a column of anion is the value of the corresponding salt.

In Table 5.1, $d\gamma/dm$ values of sodium halides, alkali metal chlorides, and two potassium halides are compared to confirm the validity of the features provided earlier. The difference between potassium halides and sodium halides is small, but KBr > KI and NaBr > NaI. The magnitude of [($d\gamma/dm$)/mN m^{-1} mol^{-1} kg] of potassium halides is specific for anions, and the difference between the values of KCl and KI is 0.44, and that between NaCl and NaI is 0.41. The minor differences in the behavior between cations and similar differences in the specific contribution between anions suggest that there is an additive rule of the ionic contribution to the surface tension increments.

Figure 5.3 schematically shows the magnitude of $d\gamma/dm$ of alkali metal halides as a function of the hydration enthalpy ΔH_h (Marcus 1987). The open circles

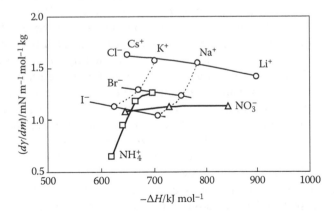

FIGURE 5.3 Variation of $d\gamma/dm$ of 1:1 electrolytes with hydration enthalpy. The data of hydration enthalpy given by Marcus (1987) are used. Salts with a common anion are connected by thin solid lines, and the salts with a common cation are connected by dotted lines. Triangles connected by a horizontal solid line are potassium, sodium, and lithium nitrates from left to right. Squares connected by a solid line represent ammonium iodide, bromide, and chloride from bottom to top.

connected by thin solid lines show the behavior of alkali metal chlorides, bromides, and iodides from top to bottom, and those connected by the dotted lines show the behavior of potassium halides and sodium halides. The grid-like lines suggest that the additive rule seems effective in the evaluation of the numerical values for the alkali metal halides, but it is important to consider whether the rule accompanies the physical significance or not. Two important features of this schematic figure are (1) the hydration of anions is an important factor in determining the increment in surface tensions, and (2) the effect of hydration of cations is less in magnitude than that of anions and a reverse in the effect of the increment in γ for alkali metal chlorides, bromides, and iodides.

5.1.2 ALKALI METAL NITRATES

When we draw the γ–m plots of the aqueous solutions of $LiNO_3$, $NaNO_3$, and KNO_3 on a figure, they are almost indistinguishable (Matubayasi and Yoshikawa 2007). The $[(d\gamma/dm)/mN\ m^{-1}\ mol^{-1}\ kg]$ values of these salts are 1.13 ($LiNO_3$), 1.13 ($NaNO_3$), and 1.08 (KNO_3). Triangles connected by a heavy solid line in Figure 5.3 illustrate that the influence of the hydration force of cations is less in magnitude than that of alkali metal chlorides. If this result is not an incidental one, there is an interaction between cations and anions in the surface region, which depends on the type of ions.

5.1.3 AMMONIUM HALIDES

The γ–m relations of the aqueous solution of ammonium halides are well represented by linear regression lines without exception, as well as those reported for alkali metal halides. In Figure 5.3, the $d\gamma/dm$ values of NH_4Cl, NH_4Br, NH_4I, and NH_4NO_3 are plotted against the hydration enthalpy of these salts at 25°C (Matubayasi et al. 2010). Since these plots can be connected by a smooth and continuous curve with positive slope, it is evident that $d\gamma/dm$ of these salts is a function of the hydration force of anions. However, the figure illustrates that the lines of these ammonium halides are not members of the lines of alkali metal halides. If the magnitude of $d\gamma/dm$ depends almost on the anions and very little on cations, the plot for NH_4Cl lies on the line connecting the alkali metal chloride instead of bromides. It is generally recognized that anions are accumulated in the upper layer of the surface region and cations are adsorbed next to the anion layer (Jarvis and Scheiman 1968, Levin 2009). The straightforward comparison between the case of ammonium and sodium halides indicates that the magnitude of the contribution to $d\gamma/dm$ depends on the anion–cation interaction as well as on the ion–water interaction. These observations for simple 1:1 electrolyte solutions reveal that the additive rule for numerical evaluation of $d\gamma/dm$ has no physical significance but is operationally effective (Marcus 2010).

5.1.4 OXOANIONS OF HALOGENS

The $\gamma-m$ plots for sodium salts of chlorates, bromates, and iodates are shown in Figure 5.4 together with that of sodium perchlorate. The shapes of the curves for these sodium halates are not as typical as expected for simple salt solutions. The plot for sodium iodate is almost linear, but the plots for $NaBrO_3$ and $NaClO_3$ are slightly concave to the concentration axis. There is an explicit description about the disproportionation of halate ions in the inorganic textbook (Cotton et al. 1995, p. 473). The disproportionation of ClO_3^- to Cl^- and ClO_4^- is thermodynamically favorable, but the reaction occurs slowly. The disproportionation of BrO_3^- is extremely unfavorable. It seems probable that the departure from a linear relation may be attributed to the disproportionation reaction in the aqueous solutions. The hydration of a halate ion that decreases in the order $IO_3^- > BrO_3^- > ClO_3^-$ will account for the order of the slopes of the $\gamma-m$ plots on the graph. These halates are pyramidal ions like NH_3, while ClO_4^- is of tetrahedral structure, which is more symmetrical than ClO_3^-. The difference in the $\gamma-m$ plots between ClO_4^- and ClO_3^- is particularly noticeable.

5.1.5 MONOBASIC PHOSPHATE ANIONS

The behavior of the polyatomic anions is difficult to explain. In contrast to the oxo anions shown in Figure 5.4, the surface tension of the solution of phosphate anions increases more rapidly than that of halate XO_3^- ions. The $\gamma-m$ plots are linear and $[(d\gamma/dm)/\text{mN m}^{-1} \text{ mol}^{-1} \text{ kg}]$ values for the aqueous solution of NaH_2PO_4 (1.85) and KH_2PO_4 (1.76) are larger than that of $NaClO_4$ (0.19), but comparable with that of NaF (1.79). The sodium and potassium monobasic phosphates have nearly the same value as well as alkali metal nitrates, while sodium and potassium dibasic phosphates have distinct slopes.

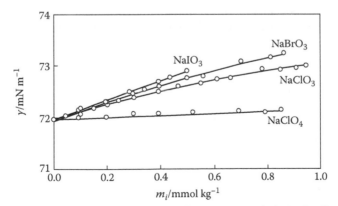

FIGURE 5.4 The variations of surface tension of the aqueous solutions of sodium chlorates, bromates, iodates, and perchlorate at 25°C.

5.1.6   Hydration Enthalpies and $d\gamma/dm$ Values of 1:1 Electrolytes

In view of the surface potential measurements for the pure water surface (Frumkin 1924, Randles 1957, Jarvis and Scheiman 1968), there are electrical double layers in terms of the asymmetric water molecules at the surface. The positive side of the dipole is oriented to the bulk water. The accumulation of anions in the surface region has been suggested from considerations of this structure and the magnitude of ion–dipole interactions. In order to obtain a better understanding of the increases in surface tension from the adsorption of ions to the air/water interface, let us now summarize the relationship between the magnitude of $d\gamma/dm$ and that of the hydration force of 1:1 salts. In Figure 5.5, the values of $d\gamma/dm$ for 12 sodium salts are plotted against the enthalpy of hydration. The values of $NaBrO_3$ and $NaClO_3$ are the slopes at the point of origin of the quadratic fit to the data without substantial reasons. Starting from the bottom left plot of sodium perchlorate, as the magnitude of hydration enthalpy of salts is increased, the $d\gamma/dm$ becomes larger and larger steeply up to the point of $NaBrO_3$. After that point, the $d\gamma/dm$ increases slowly with increasing hydration enthalpy. These sodium salts fall into two groups: those characterized by the high hydration enthalpy but insensitive to the variation of the hydration enthalpy and those that are sensitive to the variation of the hydration enthalpy. As is well known from the viscosity measurements of Jones and Dole (1929), electrolytes are classified into two groups called structure makers and structure breakers. The former ions with a positive B-coefficient are found to adsorb at the air/water interface with a hydration sheath, while the latter ions with a negative B-coefficient lose their hydration sheath at the surface (Santos et al. 2010). Since the [(B-coefficient)/ $dm^3 \ mol^{-1}$] of BrO_3^- and Cl^- are respectively 0.009 and -0.005 at 25°C (Jenkins and Marcus 1995), the $d\gamma/dm$–hydration relation reveals factors that enhance the surface tension. The distance from the image force or from the surface of discontinuity makes the $d\gamma/dm$ sensitive or insensitive to hydration enthalpy.

At the surface between air and aqueous electrolyte solutions, an electrical double layer is formed by the anions and cations. Since anions preferentially accumulate at the surface, cations cannot penetrate into the physical surface of discontinuity over the layer of anions. In view of the relationship of the distance and the double layer, it

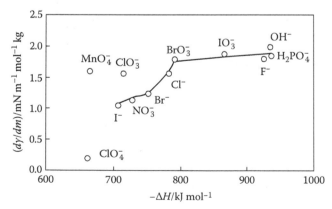

FIGURE 5.5   Comparisons of the $d\gamma/dm$–hydration enthalpy relations of 1:1 type sodium salts.

seems probable that the $d\gamma/dm$ values of a cation that are insensitive to the hydration enthalpy may be produced by the interaction with anions in the double layer.

5.2 AQUEOUS SOLUTIONS OF 2:1 ELECTROLYTES

Alkaline earth metal halides are the major 2:1 electrolytes along with those of the transition metals. The experimental γ–m plots for $MgCl_2$, $CaCl_2$, and $BaCl_2$ are shown in Figure 5.6. All three remarks listed for symmetrical 1:1 electrolytes are applicable for the data of the alkaline earth metal halides. First, the γ–m plots for all known 2:1 electrolytes are linear despite their a symmetrical valence type. Second, the γ–m plots for three alkaline earth metal chlorides have slopes similar to each other and have magnitudes of the slopes that seem reasonable for divalent ions. Finally, as we have considered for 1:1 electrolytes, the γ–m plots are sensitive to hydration enthalpy of anions. These phenomena can be illustrated graphically by plotting the values of $d\gamma/dm$ of salts against their hydration (Figure 5.7). The difference in $[\Delta H_h/\text{kJ mol}^{-1}]$ between $MgCl_2$ and $BaCl_2$ is 617 and that between $CaCl_2$ and CaI_2 is 152. On the

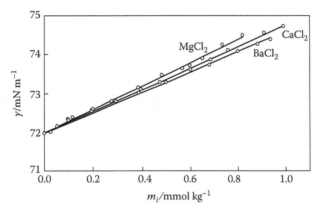

FIGURE 5.6 Surface tension–concentration curves of the aqueous solutions of magnesium, calcium, and barium chlorides at 25°C.

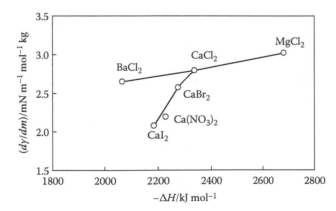

FIGURE 5.7 Variation of $d\gamma/dm$ of 2:1 electrolytes with hydration enthalpy.

other hand, the difference in $[(d\gamma/dm)/mN \, m^{-1} \, mol^{-1} \, kg]$ between $MgCl_2$ and $BaCl_2$ is 0.37 and that between $CaCl_2$ and CaI_2 is 0.72. These observations support the view derived from the surface potential measurements that there are electrical double layers in the surface region and that the electrostatic force of the cations to the physical surface of discontinuity is shielded by the layer of anions.

As shown earlier, the effect of 2:1 electrolytes on the air/water interface is similar to that presented for 1:1 electrolytes, but a positive slope of the $d\gamma/dm - \Delta H_h$ plot for a series of cations is to be noted. Figure 5.7 shows that $d\gamma/dm$ decreases in the order $MgCl_2 > CaCl_2 > BaCl_2$, which accounts for the common expectation of the relationship between the magnitude of surface tension increments and hydration enthalpies of solutes. In contrast to this, the 1:1 electrolyte in Figure 5.3 shows the reverse order to the hydration.

5.3   AQUEOUS SOLUTIONS OF 1:2 ELECTROLYTES

Most divalent anions are oxoanions. As regards the hydration force of typical divalent oxoanions, a series

$$HPO_4^{2-} > CO_3^{2-} > SO_4^{2-} > CrO_4^{2-}$$

is known (Marcus 1985, 2010). The $\gamma-m$ plots of a series of sodium salts of these anions and $d\gamma/dm - \Delta H_h$ plots of a series of alkali metal sulfates are shown in Figures 5.8 and 5.9, respectively. These figures show that remarks listed for symmetrical 1:1 electrolytes are not applicable for the data of the 1:2 electrolytes except the first one. First, each $\gamma-m$ graph can be fitted to a linear regression line. Second, the shape of curve of the $d\gamma/dm - \Delta H_h$ for alkaline metal sulfates differs from that of 1:1 or 2:1 electrolytes. It is concave upward with a minimum at

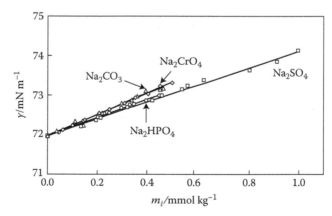

FIGURE 5.8   Surface tension–concentration curves of the aqueous solution of typical 2:1 type sodium salts: circles (Na_2HPO_4), triangles (Na_2CO_3), squares (Na_2SO_4), and diamonds (Na_2CrO_4).

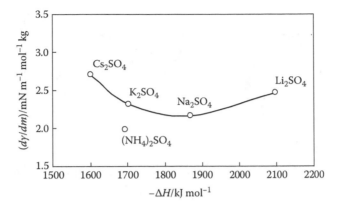

FIGURE 5.9 Variation of $d\gamma/dm$ of 1:2 electrolytes with hydration enthalpy.

sodium sulfate. The change from lithium to cesium of the chloride would increase the $[(d\gamma/dm)/\text{mN m}^{-1} \text{ mol}^{-1} \text{ kg}]$ of 1.42 to 1.62, a difference of 15%. On the other hand, the change from sodium to cesium of the sulfate would increase the value of 2.16–2.72, a difference of 25%. Finally, the characteristic of 1:2 electrolytes is that there is no relationship between the magnitude of $d\gamma/dm$ and the hydration of salts. The $d\gamma/dm$ of Na_2CO_3 and Na_2SO_4 are 2.82 and 2.16; thus the relative difference is 23% of Na_2CO_3.

5.4 AQUEOUS SOLUTIONS OF 2:2 ELECTROLYTES

Sufficient data are not available to indicate the general tendency of the 2:2 electrolytes, but the characteristic of 1:1 electrolytes will be applicable to 2:2 electrolytes. In fact, $[(d\gamma/dm)/\text{mN m}^{-1} \text{ mol}^{-1} \text{ kg}]$ values of $MgSO_4$, $CuSO_4$, and $NiSO_4$ are 1.68, 1.77, and 1.80, respectively, and are similar in magnitude to those of 1:1 electrolytes. If we suppose that the only processes leading to the elevation of surface tension are interactions between image force and ions, the $d\gamma/dm$ values of 2:2 electrolytes comparable in magnitude to 1:1 electrolytes are due to the lack of experimental cases.

5.5 HYDRATION ENTHALPIES, $d\gamma/dm$ VALUES, AND VALENCE TYPE

In the earlier sections, empirical relationships concerning the significance of hydration force and the magnitude of $d\gamma/dm$ values are considered. The application to each set of electrolytes of the same valence type is illustrated. In view of the halide anions of 1:1 and 2:1 type electrolytes, it can be pointed out that the empirical slopes, $d\gamma/dm$, are sensitive to the hydration force and are specific to each anion. Then, the $d\gamma/dm$ for 1:1 electrolytes composed of structure maker anions is insensitive to the change in the hydration of anions, although their $d\gamma/dm$

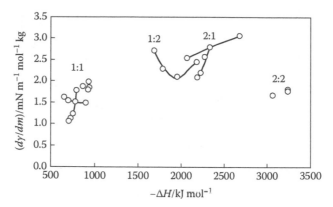

FIGURE 5.10 Comparison of $d\gamma/dm$–hydration enthalpy relations for 1:1, 2:1, 1:2, and 2:2 type electrolytes.

values are larger than those composed of structure breaker anions. Moreover, the observed empirical slopes for 1:2 electrolytes indicate no recognizable relationship between the hydration of anions and the slopes. In Figure 5.10, the $d\gamma/dm$ values of the earlier four distinctive valence types are compared in order to recognize visually how the hydration affects the surface tension increments. It appears that the $d\gamma/dm$ increases in the order 1:1 $\simeq$ 2:2 < 1:2 < 2:1 corresponding to the hydration forces of electrolytes. This figure illustrates three important points. First, it seems difficult to show a simple line on this graph that describes a single factor corresponding to the magnitude of surface tension increment. Second, two symmetrical valence types of electrolytes, 1:1 and 2:2 electrolytes, show roughly the same magnitude. The 1:2 and 2:2 electrolytes shown in the figure are all sulfates. If the anions have a decisive effect on the increment of surface tension, deviation of $MgSO_4$ from the group of 1:2 type sulfates will be too large. Finally, the 2:1 type and 1:2 type show nearly the same magnitude even though their anions are very different from each other.

The earlier three features suggest the possibility that an interesting view on surface tension increments will be found if we use the total number of moles of ions m_{ti} instead of the total number of moles of a salt m. This view is illustrated by Figure 5.11 in which $d\gamma/dm_{ti}$ values are plotted against the hydration of the electrolytes. It is found that each group is arranged almost parallel to the abscissa, and the spread of the group is very large. It seems probable from this relationship that apparent increments in surface tension of an electrolyte solution depend primarily upon the number of ions in the solution irrespective of the signs and magnitudes of charges. The graph shows regularity in one respect. The order of the halide anions depends on the hydration and is iodide < bromide < chloride. A satisfactory explanation for the order is difficult and not available. The behavior may be partially interpreted by interaction between hydrated anions and water, which depends on the structure of anions. The order of the cations of the unsymmetrical valence type may be interpreted by ionic

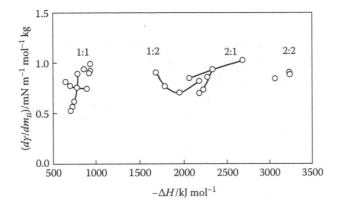

FIGURE 5.11 Comparison of $d\gamma/dm_{ti}$ relations when the total number of ions is used as a concentration variable instead of the number of moles of salts.

interactions in the double layer. The situation for the interaction in the double layer between divalent cations and univalent anions may be different from that between univalent cations and divalent anions. It may be remarked that specific differences observed in the order of cations between 1:1 and 1:2 electrolytes may also be interpreted by this specific interaction in the inhomogeneous surface region, although thermodynamics does not predict how the microscopic structure of the surface region is.

5.6 $d\gamma/dm$ VALUES OF OIL/WATER INTERFACE

Whatever the cause of the depletion of ions from the surface, the effect of ions on the oil/water interface will be helpful for understanding the ions at the interface. The surface tension measurements of the aqueous electrolyte solution against oil phase are reported by Aveyard and Saleem (1976). Their measurements revealed that the magnitudes of the $d\gamma/dm$ of air/water and dodecane/water are similar to each other and that in general those of air/water show larger values than those of dodecane/water interface. Their results are shown in Table 5.2.

More satisfying measurements of the surface tension for oil/water interface are presented by Ikeda et al. (1992) using hexane/water interface. The surface tension increment $(\gamma - \gamma_0)$ versus m plots for the aqueous solution of NaCl at the hexane/water interface are compared with those at the air/water interface in Figure 5.12. Here γ_0 represents the surface tension at air/water or oil/water interfaces with no solute present. The $[(d\gamma/dm)/mN\ m^{-1}\ mol^{-1}\ kg]$ values at the hexane/water and the air/water interfaces are 1.52 and 1.55, respectively (Figure 5.13). An equal dependence of the $\Delta\gamma$ upon concentration is very striking when considering the fact that the γ value of the air/water interface is almost 1.5 times of that the hexane/water interface at the same temperature. We will never encounter these phenomena in the adsorption of amphipathic molecules.

TABLE 5.2
[$(d\gamma/dm)$/mN m^{-1} mol^{-1} kg] Values of Some Alkali Metal Halides at the Air/Water and Dodecane/Water Interface at 20°C

Salt	Air/Water	Dodecane/Water
LiCl	1.58	1.56
NaCl	1.70	1.41
KCl	1.59	1.37
kBr	1.36	0.86
KI	1.24	−0.07
Na$_2$SO$_4$	2.66	2.40

Source: Aveyard, R. and Saleem, S.M., *J. Chem. Soc. Faraday Trans. I.*, 72, 1609, 1976.

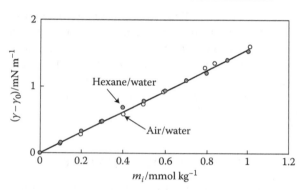

FIGURE 5.12 Comparison of the surface tension increment $(\gamma - \gamma_0)$ versus m plots for the aqueous solution of NaCl at the hexane/solution with those at the air/solution interface.

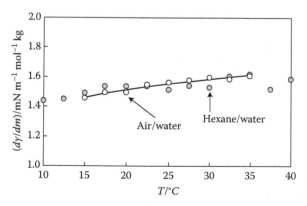

FIGURE 5.13 Variation of $d\gamma/dm$ of the aqueous solution of NaCl with temperature. Open circles and solid line are $d\gamma/dm$–T relation of hexane/water interface. Closed circles are $d\gamma/dm$ data at the hexane/water interface given by Ikeda. (From Ikeda, N., M. Aratono, and K. Motomura., *J. Colloid Interface Sci.*, 149, 208,1992.)

5.7 COMPARISON OF $d\gamma/dm$ VALUES SHOWN IN THE LITERATURES

Langmuir (1917) wrote that the data for $d\gamma/dc$ for aqueous salt solutions are rather variable. He cited six data of the aqueous solution of KCl for $d\gamma/dc$ ranges of 1.41, 1.81, 1.75, 1.75, 1.56, and 1.63 with an average of 1.65 ± 0.15. It has been tacitly assumed that lower surface tension values are contaminated, and the correct one is the higher one. The values of alkali metal halides given in the frequently cited literatures show considerable discrepancy, depending upon the researchers (Table 5.3). Pegram and Record (2007) surveyed the literature and reported the average values of the electrolytes that are shown in the last column of the table. The average value from the first to the fourth column of KCl is 1.59 ± 0.05, which is very close to the average value of the data referred by Langmuir. This indicates definitely that for the surface tension measurements, surface active contaminations are not a major factor for the scattering data, but such an assumption may be one of the factors. Further, considering the three significant figures of $d\gamma/dm$, the values for KCl in the table suggest that the four authors reported the same value and their measurements would give equally the same precision provided that they use, the same density data, and so on. Data given in the fifth column in the table are larger than those of the other authors, but it is hard to review them because of the lack of the necessary temperature data in equilibrium.

TABLE 5.3
Comparison of [($d\gamma/dm$)/mN m^{-1} mol^{-1} kg] Values of Alkali Metal Chlorides and Sodium Halides at 25°C

Salt	Matubayasi	Heydweiller	Aveyard[a]	Johnasson	Weissenbor[b]	Pegram[c]
LiCl	1.42	2.11	1.58	—	1.98	1.65
NaCl	1.55	1.58	1.70	1.75	2.08	1.73
KCl	1.57	1.54	1.59	1.65	1.85	1.59
CsCl	1.63	—	—	—	1.52	1.57
NaF	1.79	—	—	—	1.83	1.81
NaBr	1.55	1.27	1.57	—	1.83	1.47
NaI	1.04	—	—	1.21	1.23	1.14

Source: Data are taken from Heydweiller, A., *Ann. Physik.* [4], 33, 145, 1910; Aveyard, R. and Saleem, S.M., *J. Chem. Soc. Faraday Trans. I.*, 72, 1609, 1976; Johnasson, J. and Eriksson, C.K., *J. Colloid Interface Sci.*, 49, 469, 1974; Weissenborn, P.K. and Pugh, R.J., *J. Colloid Interface Sci.*, 184, 550, 1996; Pegram, L.M. and Record, Jr., M.T., *J. Phys. Chem. B.*, 111, 5411, 2007.

[a] Values at 20°C.

[b] Temperature is not given provided that the temperature dependence is smaller than the difference between different studies.

[c] Average value of $d\gamma/dm$ compiled from literature under the assumption that small systematic differences between different studies are more significant than differences in temperature.

REFERENCES

Aveyard, R. and S. M. Saleem. 1976. Interfacial tensions at alkane-aqueous electrolyte interfaces. *J. Chem. Soc. Faraday Trans. I.* 72: 1609–1617.

Cotton, F. A., G. Wilkinson, and P. L. Gaus. 1995. *Inorganic Chemistry*, 3rd edn. Wiley, New York.

Frumkin, A. Z. 1924. Phasengrenzkräfte und adsorption an der trennungsfläche luft Lösung anorganisheher electrolyte. *Z. Phys. Chem.* 109: 34–48.

Heydweiller, A. 1910. Über physikalische eigenshaften von lösungen in ihrem zusammenhang. II Overflächenspannung und elektrisches leitvermögen wässeriger salzlösungen. *Ann. Physik.* [4] 33: 145–185.

Ikeda, N., M. Aratono, and K. Motomura. 1992. Thermodynamic study on the adsorption of sodium chloride at the water/hexane interface. *J. Colloid Interface Sci.* 149: 208–215.

Jarvis, N. L. and M. A. Scheiman. 1968. Surface potentials of aqueous electrolyte solutions. *J. Phys. Chem.* 72: 74–78.

Jenkins, H. D. B. and Y. Marcus. 1995. Viscosity B-coefficients of ions in solution. *Chem. Rev.* 95: 2695–2724.

Johansson, J. and C. K. Eriksson. 1974. γ and $d\gamma/dT$ measurements on aqueous solutions of 1, 1 – electrolytes. *J. Colloid Interface Sci.* 49: 469–480.

Jones, G. and M. Dole. 1929. The viscosity of aqueous solutions of strong electrolytes with special reference to barium chloride. *J. Am. Chem. Soc.* 51: 2950–2964.

Langmuir, I. 1917. Fundamental properties of solids and liquids II. *J. Am. Chem. Soc.* 39: 1848–1906.

Levin, Y. 2009. Polarizable ions at interfaces. *Phys. Rev. Lett.* 102: 147803.

Marcus, Y. 1985. *Ion Solvation*. John Wiley & Sons, New York.

Marcus, Y. 1987. The thermodynamics of solvation of ions. Part 2. The enthalpy of hydration at 298.15 K. *J. Chem. Soc. Faraday Trans. 1.* 83: 339–349.

Marcus, Y. 2010. Surface tension of aqueous electrolytes and ions. *J. Chem. Eng. Data* 55: 3641–3644.

Matubayasi, N., K. Takayama, and T. Ohata. 2010. Thermodynamic quantities of surface formation of aqueous electrolyte solutions. IX. Aqueous solutions of ammonium salts. *J. Colloid Interface Sci.* 344: 209–213.

Matubayasi, N., K. Tsunetomo, I. Sato, R. Akizuki, T. Morishita, A. Matuzawa, and Y. Natsukari. 2001. Thermodynamic quantities of surface formation of aqueous electrolyte solutions IV. Sodium halides, anion mixtures, and sea water. *J. Colloid Interface Sci.* 243: 444–456.

Matubayasi, N., K. Yamamoto, S. Yamaguchi, H. Matsuo, and N. Ikeda. 1999. Thermodynamic quantities of surface formation of aqueous electrolyte solutions III. Aqueous solutions of alkali metal chloride. *J. Colloid Interface Sci.* 214: 101–105.

Matubayasi, N. and R. Yoshikawa. 2007. Thermodynamic quantities of surface formation of aqueous electrolyte solutions VII. Aqueous solution of alkali metal nitrates $LiNO_3$, $NaNO_3$, and KNO_3. *J. Colloid Interface Sci.* 315: 597–600.

Pegram, L. M. and M. T. Record Jr. 2007. Hofmeister salt effects on surface tension arise from partitioning of anions and cations between bulk water and the air-water interface. *J. Phys. Chem. B.* 111: 5411–5417.

Randles, J. E. B. 1957. Ionic hydration and the surface potential of aqueous electrolytes. *Discuss. Faraday Soc.* 24: 194–199.

Santos, A. P. dos, A. Diehl, and Y. Levin. 2010. Surface tensions, surface potentials, and the Hofmeister series of electrolyte solutions. *Langmuir* 26: 10778–10783.

Weissenborn, P. K. and R. J. Pugh. 1996. Surface tension of aqueous solutions of electrolytes: Relationship with ion hydration, oxygen solubility, and bubble coalescence. *J. Colloid Interface Sci.* 184: 550–563.

6 Adsorption of Ions at Air/Water Interface

In the preceding chapter, the surface tension and the $d\gamma/dm$ were considered. Many of the works for simple electrolyte solutions thus far have aimed to consider how and why the $d\gamma/dm$ has positive values. In general, a practical application of surface thermodynamics to the experimental γ–m relations is carried out to obtain information about the surface excess densities of solute molecules in the surface region. At a given temperature and pressure, the surface tension varies according to the relationship $d\gamma = \Gamma_i^H d \ln \mu_i$. We know intuitively from common experience that the surface tension should decrease and the adsorption will be increased because the chemical potential increases as the concentration increases But for the positive change in the surface tension, we have less common experience about the deficiency of solute in the surface region. In this chapter, we will consider the information available from the surface excess densities of ions. Section 6.1 is devoted to the presentation of Γ_i^H of 1:1, 1:2, and 2:1 electrolytes. The Γ_i^H value is relatively insensitive to variations in ionic species. Then we consider the variation of Γ_i^H / m with concentration in order to see the relation between surface activity and ionic species.

6.1 SURFACE EXCESS DENSITY OF IONS

6.1.1 Aqueous Solutions of 1:1 Electrolytes

A positive surface excess density, in general, has no uncertainties in the prediction that Γ_i^H is almost the same with the concentration in the surface region and the negative surface excess density of electrolyte delivered the idea of the "ion-free layer" in a similar manner of positive adsorption. The difference in the negative surface excess density between electrolytes has been assigned to the difference in the thickness of the ion-free layer. In considering properties of various adsorbed films of surfactants, it is common practice to plot the adsorption as a function of concentration if the shape of the curve gives the Langmuir adsorption isotherm. However, such a plot for adsorption from simple electrolyte solutions may not be common practice in the literatures. The surface excess density of simple electrolyte i, Γ_i^H, however, delivers some message because this is a quantity in the first place derived from the surface tension measurements according to the Gibbs surface thermodynamics.

Typical plots for sodium halides are shown in Figure 6.1. The bottom curve of the figure represents the behavior of the sodium fluoride, and the thin dotted line next to the curve is Γ_i^H without the correction of the activity coefficient. It is clear that both curves vary almost linearly with the concentration and that the surface deficiency of

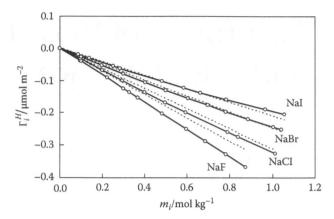

FIGURE 6.1   Adsorption of sodium halides at air/solution interface at 25°C.

sodium fluoride in the surface region is remarkable. The deficiency decreases in the order NaF > NaCl > NaBr > NaI, and the hydration enthalpy of these salts decreases in the same order. There is one important point to be noted here. The difference in the deficiency Γ_i^H of two salts is equivalent to the difference in the surface density, Γ_i^I, of the two salts in the surface region. We have no means of obtaining the number of moles of salts in the surface region experimentally, but the plots given in the figure suggest that there are appreciable amounts of NaI in the surface region when compared with that of NaF. For example, the $\left[\Gamma_i^I/\mu\text{mol m}^{-2}\right]$ of NaI is larger by 0.038 than that of NaBr and by 0.174 than that of NaF at 0.8 mol kg^{-1}, respectively. There may be considerable amount of ions in the surface region, and it is true that Γ_i^I increases almost linearly with increasing bulk concentration of the electrolytes.

From Equation 2.137, we have

$$\Gamma_i^H = \Gamma_+^H = \Gamma_-^H$$

for 1:1 electrolytes. Then we can replace the label on the vertical axis with either Γ_+^H or Γ_-^H without any modification of the graph. If we change the label from Γ_i^H to $\Gamma_{\text{Na}^+}^H$, the figure represents the variation of the surface excess density of sodium ion with increasing concentration. The graph of sodium ion shows that the adsorption of sodium ion depends largely upon the co-ion species.

In the previous chapter, consideration of the $d\gamma/dm$ shows that the effect of the cation species is less significant than that of anions. Then, a graph of the adsorption for the effect of the cation species is illustrated in Figure 6.2. The solid lines represent the values for LiCl, NaCl, KCl, and CsCl from top to bottom, respectively, and the thin dotted lines are those obtained assuming ideal dilute solutions. The discrepancy between the sets of dotted lines is small, but those solid lines are appreciable. Again, if we change the label of the vertical axis from Γ_i^H to $\Gamma_{\text{Cl}^-}^H$, the graph shows that the adsorption of Cl$^-$ ions to the surface depends considerably upon the cation species. The adsorption of Cl$^-$ ions with Li$^+$ is larger by 0.10 μmol m^{-2} at 0.8 mol kg^{-1} than that with Cs$^+$. These observations indicate that

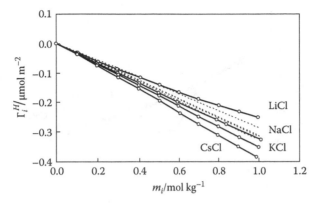

FIGURE 6.2 Adsorption of alkali metal chlorides at air/solution interface at 25°C.

the adsorption of ions changes its value as a result of ionic interactions between anions and cations in the surface region and that additive properties of ions in the $d\gamma/dm$ for alkali metal halide are an incidental phenomenon.

6.1.2 ELECTROLYTES OF UNSYMMETRICAL VALENCE TYPE

There have been few systematic measurements of the surface tension of aqueous electrolyte solutions, so it would be meaningful to illustrate the adsorption of the electrolytes of the unsymmetrical valence type. The surface excess densities for series of 2:1 electrolytes, calcium halides ($CaCl_2$, $CaBr_2$, CaI_2), and alkaline earth metal chlorides ($MgCl_2$, $CaCl_2$, $BeCl_2$) are shown in Figures 6.3 and 6.4, respectively. The $\left[\Gamma_i^H \mu\text{mol m}^{-2}\right]$ values of $CaCl_2$ decrease to about −0.3 at 1 mol kg^{-1}, which is almost equal to those of 1:1 electrolytes. The adsorption of CaI_2 is larger than that of $CaCl_2$ by about 0.10 μmol m^{-2} when compared at the concentration of 0.8 mol kg^{-1}. There is a similarity in the order and magnitude of the effect of halide anions between 1:1 and 2:1 electrolytes. However, Figures 6.1 through 6.4 show substantial differences

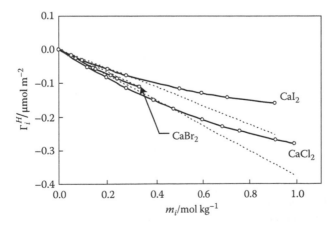

FIGURE 6.3 Adsorption of calcium halides at air/solution interface at 25°C.

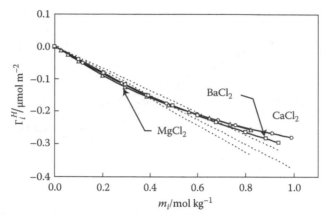

FIGURE 6.4 Adsorption of magnesium, calcium, and barium chlorides at air/solution interface at 25°C.

between them. First, the adsorption isotherms of the electrolytes of the unsymmetrical valence type are not a linear function of concentration but concave upward. The calculated Γ_i^H values are not in accord with the linear dotted lines and depend on the term of the activity coefficients in the bulk phase. Then the graphs of the three isotherms for $MgCl_2$, $CaCl_2$, and $BeCl_2$ are almost overlapping one another. It should be noted that the adsorption is determined by the valence type of the electrolytes and not by the properties of the divalent cations.

The 1:2 electrolytes also show adsorption isotherms different from those of 1:1 electrolytes, though the variation of 1:2 electrolytes is limited. The solid and dotted lines in Figure 6.5 represent the sulfates and hydrogen phosphates, respectively. The open circles and triangles show sodium and potassium salts, respectively. The four curves in the figure represent the difficulty of finding any relationship between the

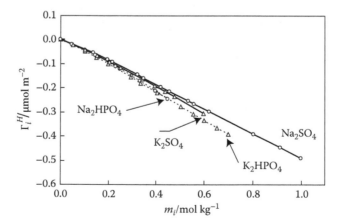

FIGURE 6.5 Adsorption of 1:2 electrolytes: sodium sulfate, sodium hydrogen phosphate, potassium sulfate, and potassium hydrogen phosphate at 25°C. The solid and dotted lines represent the sulfates and hydrogen phosphates, respectively. The open circle and triangle show sodium and potassium salts, respectively.

types of anions and cations needed to suggest distinctive adsorption isotherms. The adsorption, Γ_i^H, versus concentration curves are insensitive to the changes in the variation of ions and ion pairs.

6.2 Γ^H/m OF ELECTROLYTES

A molecule that lowers the surface tension of the solution is called a surface active substance, and the measure of the lowering is called surface activity. In Section 4.3, we mentioned briefly the definition of the surface activity of surface active substances along with Equation 4.22:

$$\frac{\Gamma_i^H}{m_i} = -\frac{1}{RT}\left(\frac{\partial \gamma}{\partial m_i}\right)_{T,p}. \tag{4.22}$$

In general, the slopes of γ–concentration relations at infinite dilution are used as a measure of the surface activity, and sometimes Γ_i^H values at the fixed concentration are also used. Consideration of an attempt to explain the specific distribution of ions between surface region and the bulk solution leads to the concept of the surface activity of ions, although the definition of it remains unclear even when considering the surfactant molecules. For an electrolyte i that dissociates into ν_+ cations and ν_- anions, we can write

$$\frac{\Gamma_i^H}{m_i} = -\frac{1}{\left(\nu_+ + \nu_-\right)RT}\left(\frac{\partial \gamma}{\partial m_i}\right)_{T,p} \bigg/ \left[1+\left(\frac{\partial \ln f_\pm}{\partial \ln m_\pm}\right)_{T,p}\right]. \tag{6.1}$$

When we use the molar concentration c_i instead of the molal concentration m_i, the left side of the equation gives τ, the thickness of a single layer of molecules of the solvent denoted by Langmuir (1917, Eq. 31). According to Gibbs' article (Gibbs, Eq. 515), the superficial density of component i in reference to the dividing surface (surface of tension) is given by

$$\Gamma_i = \frac{n_i}{A} - l^\alpha c_i^\alpha - l^\beta c_i^\beta.$$

If we define a dividing surface that would make $\Gamma_i = 0$, then

$$\Gamma_i = 0 = \frac{n_i}{A} - (l^\alpha + \tau)c_i^\alpha - (l^\beta - \tau)c_i^\beta.$$

The distance between the two dividing surfaces τ is given as

$$\tau = \frac{\Gamma_i}{c_i^\alpha - c_i^\beta}.$$

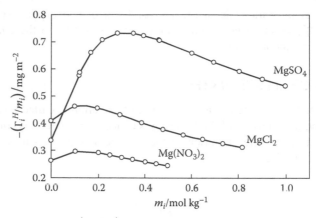

FIGURE 6.6   Variation of $-\left(\Gamma_i^H/m_i\right)$ of magnesium sulfate, chloride, and nitrate solutions as a function of concentration at 25°C.

When $c_i^\beta = 0$, Γ_i/c_i^α represents the relative distance between the two dividing surfaces. Of course, Γ_i^H/m_i is not the same that they intended to convey by Γ_i/c_i^α. Nevertheless, it is more appropriate to use this ratio Γ_i^H/m_i than Γ_i^H or $(\partial\gamma/\partial m)_{T,p}$ for the quantitative index of the surface activity of an electrolyte or an ion, because the concept of the surface activity refers to an equilibrium distribution of a solute between the bulk solution and the surface region. This is most plainly seen in a $-\left(\Gamma_i^H/m_i\right)$ versus m_i relation. An example of such a graph is shown in Figure 6.6, where $MgSO_4$, $MgCl_2$, and $Mg(NO_3)_2$ are compared. Since we have plotted $-\Gamma_i^H/m_i$, the larger the value, the smaller is the surface activity. The plots of solutions at infinite dilution are obtained by setting the derivative of the activity coefficient with respect to m_i equal to zero. The plots increase sharply at low concentrations in the figure as the activity coefficient of the salts steeply decreases, passes through a maximum at about 0.2 mol kg^{-1}, and decreases gradually with concentration. Since the drops in the activity coefficient of 2:2 electrolytes are larger than those of 1:2 electrolytes, the increases of the plots of magnesium sulfate solutions are remarkable. These observations indicate that the surface activity of magnesium salts increases in the order $MgSO_4 < MgCl_2 < Mg(NO_3)_2$ at the higher concentration range of the figure. It is clearly seen that the graph has the advantages that (1) the plots on the vertical axis follow the order in the magnitude of $d\gamma/dm$; (2) it is explicit that the value depends significantly on the concentration and type of electrolyte, which are not shown clearly on the adsorption isotherms; and (3) it is possible to make plots of ions instead of electrolytes and of the interdependence of pairs of anions and cations. The label of the vertical and horizontal axes in Figure 6.6 can be changed to Γ_{Mg}^H/m_{Mg} and m_{Mg} without any modification to the figure. We see that surface deficiency of Mg^{2+} is significantly dependent upon the type of co-ions and concentration.

6.2.1   SODIUM HALIDES

In general, the $-\left(\Gamma_i^H/m_i\right)$ value first increases with concentration, immediately reaches the maximum, and thereafter decreases very slowly with concentration.

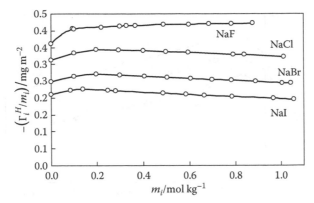

FIGURE 6.7 Variation of $-\left(\Gamma_i^H/m_i\right)$ of sodium halides as a function of concentration at 25°C.

The shapes for sodium halides are compared in Figure 6.7. The top curve for NaF does not show the maximum, but the other three sodium halides show the typical shape. However, the variations of $-\left(\Gamma_i^H/m_i\right)$ with concentration after the maximum are very slow. This tendency is observed more clearly for solutions of NH_4Cl, NH_4Br, NH_4NO_3, and NH_4I. The curves are almost flat for solutions of more than 0.1 mol kg^{-1} and are same as the curve of NaF. These observations suggest that the concentrations of these 1:1 electrolytes in the surface region increase almost linearly with increasing bulk concentrations. The figure also shows the variation of $-\left(\Gamma_X^H/m_X\right)$ of halide anions, and also the variation $-\left(\Gamma_{Na}^H/m_{Na}\right)$ of sodium ion, which depends on its co-ion. These explanations are based on the assumption that the principle of electro-neutrality should be held in the whole system, in the bulk phase, and in the surface region when the system is in equilibrium. It is to be noted that the surface deficiency of anion decreases in the order $F^- > Cl^- > Br^- > I^-$.

6.2.2 CALCIUM HALIDES

The $-\left(\Gamma_i^H/m_i\right)$ values of calcium halides show pronounced maximums in the concentration range of 0.1–0.2 mol kg^{-1} and thereafter decrease steeply with concentration (Figure 6.8). The deficiency of the ions is remarkable at a concentration of around 0.1 mol kg^{-1}, but thereafter the adsorption of ions becomes pronounced with increasing concentration. The graph shows a roughly parallel family of curves, indicating that the magnitude of the deficiency of ions is primarily concerned with the halide anions. Further, the maximum and the significant lowering of $-\left(\Gamma_i^H/m_i\right)$ may be assigned to the interaction between the divalent cation and the halide ions, since the lowering just described is not observed for sodium halide systems.

6.2.3 ALKALI METAL CHLORIDES, NITRATES, AND DIHYDROGEN PHOSPHATES

Figure 6.9 shows plots of $-\left(\Gamma_i^H/m_i\right)$ against concentration for a series of alkali metal chloride and nitrate solutions. The $-\left(\Gamma_i^H/m_i\right)$ values of infinitely dilute solutions

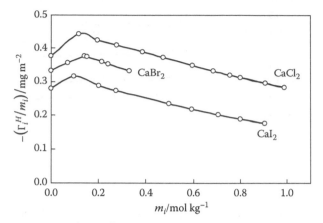

FIGURE 6.8 Variation of $-\left(\Gamma_i^H / m_i\right)$ of calcium halides as a function of concentration at 25°C.

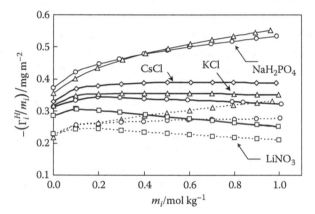

FIGURE 6.9 Variation of $-\left(\Gamma_i^H / m_i\right)$ of alkali metal chlorides, nitrates, and dihydrogen phosphates at 25°C. Squares, circles, triangles, and diamond show lithium, sodium, potassium, and cesium, respectively. Heavy solid line, thin solid line, and dotted line show chlorides, dihydrogen phosphates, and nitrates, respectively.

of $LiNO_3$, $NaNO_3$, and KNO_3 are almost identical, and those of $LiCl$, $NaCl$, KCl, and $CsCl$ show the same tendency. However, their differences become noticeable and appreciable with increasing concentration as shown in the figure. It can be seen from this figure that the hydration of cation that decreases in the order $Li^+ > Na^+ > K^+ > Cs^+$ will not account for the order of the adsorption into the surface region. The surface deficiency of cation decreases in the order $Cs^+ > K^+ > Na^+ > Li^+$. The curves of NaH_2PO_4 and KH_2PO_4, which consist of the structure-making anion, are also plotted in the figure. These two overlapping curves of sodium and potassium salts show two distinctive properties from other 1:1 electrolytes. First, these curves are associated with a large surface deficiency, and then the deficiency increases steadily with increasing concentration.

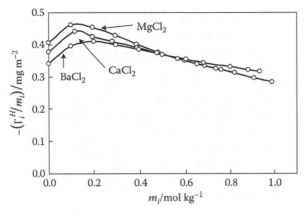

FIGURE 6.10 Variation of $-\left(\Gamma_i^H / m_i\right)$ of 2:1 electrolytes as a function of concentration at 25°C.

6.2.4 ALKALINE EARTH METAL CHLORIDES

A number of $-\left(\Gamma_i^H / m_i\right)$–$m$ curves suggest the generalization that $-\left(\Gamma_i^H / m_i\right)$ for divalent cations passes through the large maximum at a concentration of around 0.1 mol kg^{-1} (Figure 6.10). After the peak, the plots of magnesium, calcium, and barium chlorides lie on a line. On account of the earlier observations, we will assume that the interaction between the divalent cation and the two univalent anions plays a significant contribution to the deficiency of ions.

6.2.5 AQUEOUS SOLUTION OF 1:2 ELECTROLYTES

In Figure 6.11, the $-\left(\Gamma_i^H / m_i\right)$ values are plotted against concentration for the aqueous solutions of Na_2SO_4, K_2SO_4, Na_2HPO_4, and K_2HPO_4, whose anions are

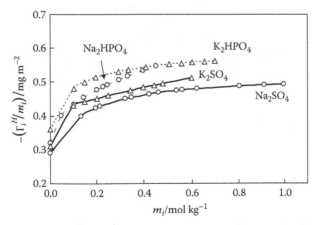

FIGURE 6.11 Variation of $-\left(\Gamma_i^H / m_i\right)$ of 1:2 electrolytes as a function of concentration at 25°C.

structure makers. It is observed that $-\left(\Gamma_i^H/m_i\right)$ increase steadily as just observed for NaH_2PO_4. There are more data for divalent anions such as CrO_4^{2-} and CO_3^{2-}, but the variation in the salt shows no pronounced effect on the shapes of the curves. It seems probable that there is no simple factor on which the order of the curve depends. For example, we find it difficult to explain the difference between the behaviors of 1:2 and 2:1 electrolytes. To avoid confusion in the figure, the curves for Li_2SO_4 and Cs_2SO_4 are omitted. The values for Cs_2SO_4 increase progressively at low concentration above that for other sulfates, suggesting that the surface deficiency is larger than others. At the higher concentration region of this figure, the surface deficiency decreases in the order $Cs_2SO_4 > K_2SO_4 > Na_2SO_4 > Li_2SO_4$ in a similar manner observed for alkali metal nitrates and chlorides. Thermodynamics does not explain this order clearly.

6.2.6 Aqueous Solution of 2:2 Electrolytes

In Figure 6.12, the $-\left(\Gamma_i^H/m_i\right)$ values of $MgSO_4$, $CuSO_4$, and $NiSO_4$ are plotted against concentration. Each of the curves is characterized by a large surface deficiency of ions and decreases in deficiency with increasing concentration after the maximum. It was pointed out earlier that the structure maker anions determine the shape of $-\left(\Gamma_i^H/m_i\right)$ versus concentration curves so that the $-\left(\Gamma_i^H/m_i\right)$ value increases steadily with concentration. Further, it was also pointed out that the 2:1 electrolyte system is unusual in decreasing its surface deficiency with increasing concentration after the large maximum regardless of anion species. The observation for 2:2 electrolytes of symmetrical valence type suggests that there is no relation between the maximum and the unsymmetrical valence type of the electrolyte, but the divalent cation is significant.

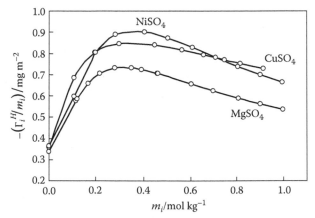

FIGURE 6.12 Variation of $-\left(\Gamma_i^H/m_i\right)$ of 2:2 electrolytes as a function of concentration at 25°C.

6.3 Γ_k^H/m_k OF IONS

For an electrolyte i that dissociates into ν_+ cations and ν_- anions, the molality of the cations and anions are $m_+ = \nu_+ m_i$ and $m_- = \nu_- m_i$, respectively. Similarly, the surface excess densities of cations and anions are $\Gamma_+^H = \nu_+ \Gamma_i^H$ and $\Gamma_-^H = \nu_- \Gamma_i^H$, respectively. Then we have

$$\frac{\Gamma_i^H}{m_i} = \frac{\Gamma_+^H}{m_+} = \frac{\Gamma_-^H}{m_-} \tag{6.2}$$

for all valence types of the electrolytes. In this section, we present some useful examples for individual ions.

6.3.1 IODINE IONS

When we consider the adsorption of ions, we have been influenced strongly by the fact that the anions are first accumulated near the physical surface of discontinuity and then the next layer is formed by the adsorbed cations. For the aqueous solutions of alkali metal halides, further, it is an experimental fact that the surface tension of the solution behaves as if there is an additive rule for the contribution of ionic species. Let us compare $-\left(\Gamma_{I^-}^H/m_{I^-}\right)$ of iodine ion. The value of $-\left(\Gamma_{I^-}^H/m_{I^-}\right)$ is plotted against the concentration of iodine ion in Figure 6.13 as an example of the behavior of a univalent anion. The figure illustrates a meaningful behavior of anions by comparison of the curves. The value of $-\left(\Gamma_{I^-}^H/m_{I^-}\right)$ clearly depends to a high degree upon the co-ion species. It clearly follows from the surface potential measurement suggesting the formation of electrical double layer at the surface. The formation of the double layer from ions of an electrolyte suggests that there is an interaction between the two layers. The meaningful behavior is the fact that the surface tension measurements have detected such an interaction between layers of ions in the surface

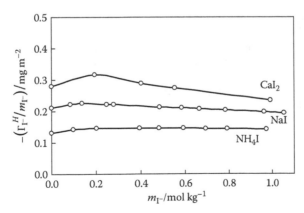

FIGURE 6.13 The effect of co-ion on the adsorption of iodine anion.

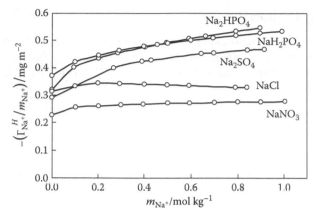

FIGURE 6.14 The effect of co-ion on the adsorption of sodium cation.

region where available moles of ions may be very limited. It is easy to see in the figure how divalent calcium cations enhance the surface deficiency of iodine anions.

6.3.2 Sodium Ions

Sodium ions are commonly studied ions. The surface deficiency of sodium ions is much dependent on the co-ion species, and the divalent anions enhance the surface deficiency more effectively than the univalent anions (Figure 6.14). However, there is a remarkable similarity between the $-\Gamma_{Na^+}^H / m_{Na^+}$ versus m_{Na^+} curves of 1:2 Na_2HPO_4 and 1:1 NaH_2PO_4. In the absence of sufficient experimental data, it is hard to assign specific values to these electrolytes, but the large surface deficiency of 1:1 NaH_2PO_4 can be attributed to the hydration of structure maker univalent anion. The mutual dependence of anions and cations in the surface region as shown in Figures 6.13 and 6.14 suggests that the additive property will not be hold for the $d\gamma/dm$ and that the interaction between anions and cations in the surface region is significant.

REFERENCES

Gibbs, J. W. 1993. Influences of discontinuity upon the equilibrium of heterogeneous masses—Theory of capillarity. In *Collected Works*, Vol. 1. pp. 219–311, Ox Bow Press, Woodbridge, CT.

Langmuir, I. 1917. Fundamental properties of solid and liquid II. *J. Am. Chem. Soc.* 39: 1848–1906.

7 Surface Tension of Solutions and Temperature

At the beginning of the study of the electrolyte at air/water interface, Langmuir (1917), Harkins and McLaughlin (1925), and Harkins and Gilbert (1926) had assumed the ion-free layer in the surface region as an operational parameter to distinguish a contribution of an individual electrolyte to the surface. From the thermodynamic viewpoint in which extensive variables are considered as an average of microscopic quantities, a horizontally isotropic surface region has laterally a macroscopic dimension, but the thickness of several layers is a negligible one. Gibbs, accordingly, introduced the surface excess quantities and successfully described the quantities inherent to the surface region. In the preceding chapters, we discussed the adsorption and deficiency of the simple electrolytes based on the consideration of surface excess densities of the aqueous solutions. Systematic measurements of the $\gamma-m$ relations and fundamental consideration for the electrolyte solutions are presented. In this chapter, we will consider the entropy changes associated with the adsorption of electrolyte. Since an electrical double layer formed by anions and cations in the surface region is confirmed by measurements of the electric potential (Frumkin 1924, Randles 1957, Jarvis and Scheiman 1968), we are interested in how thermodynamic quantities will be affected by such structures formed in the surface region. We believe that consistent consideration of the surface excess quantities will be useful for understanding the surface region without unnecessary confusion.

The entropy changes with the adsorption will be introduced with the experiments of the surface tension as a function of temperature:

$$\Delta s = -\left(\frac{\partial \gamma}{\partial T}\right)_{p,m_i}.$$

As an example, the $\gamma-T$ plots of sodium iodide solutions are illustrated in Figure 7.1. As has already been pointed out, the $\gamma-T$ plot of the pure water surface is slightly concave to the temperature axis, it is to be expected that the $\gamma-T$ plots for the solution also show slightly curved plots concave to the temperature axis. However, we see that the experimental curves can be treated as a strictly linear line between 15°C and 35°C and that the estimated slopes of linear regression lines are clearly a function of the concentration. The variation will also be a function of surface excess densities of electrolytes. The slope can be evaluated with three significant figures, but the variation of the slope with concentration is generally less than 10% of that of

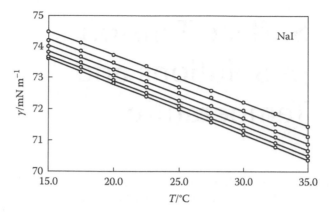

FIGURE 7.1   Surface tension–temperature curves for the aqueous solution of sodium iodide. Molalities of solutions are 0.0, 0.1380, 0.2815, 0.4846, 0.6643, and 0.9654 mol kg^{-1} from bottom to top.

pure water. Many examples displayed in this chapter will make clear the properties in the surface region, yet the significant figures of the values derived from the slope of γ–T relations are limited, and the surface tension measurements are accompanied by relatively significant accidental errors.

7.1 $d\gamma/dm$ VALUES AND TEMPERATURE

The fact that the $d\gamma/dm$ values can be used as a characteristic variable of electrolyte solutions is due to the strictly linear relationship between surface tension and concentration. It is empirically found that, in general, $d\gamma/dm$ is independent of concentration but increases with increasing temperature. The variation of $d\gamma/dm$ of sodium iodide solutions with temperature is shown in Figure 7.2. The dotted and solid lines in the figure illustrate a linear regression line and a quadratic curve, respectively. It is seen that the peculiarities of this curve are the positive slope and the slightly curved line.

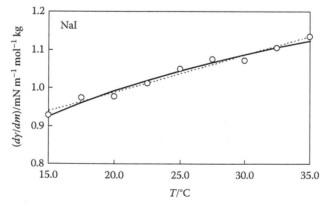

FIGURE 7.2   Variation of $d\gamma/dm$ of sodium iodide solutions with temperature. The dotted and solid lines are the linear and quadratic regression lines, respectively.

Since the surface tension is a thermodynamic quantity, the following reciprocity relationship should be held:

$$\left[\frac{d}{dT}\left(\frac{d\gamma}{dm}\right)_{T,p}\right]_{p,m} = \left[\frac{d}{dm}\left(\frac{d\gamma}{dT}\right)_{p,m}\right]_{T,p}. \tag{7.1}$$

This relation suggests two important factors in considering how experimental surface tension and related quantities vary with temperature and concentration.

First, when γ is strictly a linear function of m,

$$\gamma = \gamma_m(T)m + \gamma_{m0}(T),$$

where $\gamma_m(T)$ and $\gamma_{m0}(T)$ are the two constants of the linear equation at a fixed temperature. Substitution of this linear relation into the left side of the preceding Equation 7.1 yields

$$\frac{d}{dT}\left(\frac{d\gamma}{dm}\right)_{T,p} = \frac{d}{dT}\gamma_m(T) = \gamma'_m(T). \tag{7.2}$$

Since the right side of Equation 7.1 is equal to $-(d\Delta s/dm)_{T,p}$, we see that the linear γ–m relation leads inevitably to the linear Δs–m relationship at the fixed temperature.

In the same way, if we assume that surface tension is a linear function of temperature,

$$\gamma = \gamma_T(m)T + \gamma_{T0}(m).$$

Under this condition, the right side of Equation 7.1 must have a fixed value at a given concentration, since

$$\frac{d}{dm}\left(\frac{d\gamma}{dT}\right)_{p,m} = \frac{d}{dm}\gamma_T(m) = \gamma'_T(m). \tag{7.3}$$

Since Equation 7.1 gives

$$\frac{d}{dT}\left(\frac{d\gamma}{dm}\right) = \gamma'_T(m), \tag{7.4}$$

we see again that the derivative of γ with respect to the concentration must be a linear function of temperature at a given concentration.

The earlier considerations suggest that the linear γ–m relations and a slightly curved line shown in Figure 7.2 are conflicting observations. In view of the lack of an exact theoretical relation in the moderately concentrated solutions here, it is difficult to know which of the curves represent the properties of solutions. It is probable that the $\gamma-m$ and $\gamma-T$ relationships of the aqueous solutions are not strictly linear relations, yet the curvature cannot be described mathematically because of the limited significant figures of the experimental surface tension and its derivatives.

7.2 $d\gamma/dm$ AND Δs OF ELECTROLYTE SOLUTIONS

7.2.1 ALKALI METAL CHLORIDES

The deduction in the preceding section has shown that the entropy change of adsorption can be obtained in two processes using Equations 2.95 and 7.1, respectively. For most solutions, the $d\gamma/dm$ may not be a linear function of the temperature, but it increases almost linearly with increasing temperature as illustrated in Figure 7.2. The slope of the $d\gamma/dm$ versus temperature curve is almost equal to the slope, $-(d\Delta s/dm)_{T,p}$. In Figure 7.3, the $d\gamma/dm$ values of alkali metal chlorides are plotted against temperature. A roughly parallel family of $d\gamma/dm-T$ curves is obtained although the value of $d\gamma/dm$ itself is distinctive for each salt. It is found that the curves are arranged in the order Cs > K = Na > Li over the temperature range observed. The difference in the value of $d\gamma/dm$ between 15°C and 35°C is approximately 0.19 mN m^{-1} mol^{-1} kg for lithium chlorides. The value suggests that the $d\gamma/dm$ value increases evidently more than the error range, provided that the experimental error may appear in the second decimal point. These positive increases in $d\gamma/dm$ correspond exactly to the decrease in the entropy changes of the adsorption of electrolytes in the surface region. The slopes of the linear regression lines are shown in Table 7.1.

The Δs values evaluated from the slopes of linear least-square lines of $\gamma-T$ curves are plotted against concentration in Figure 7.4. If we keep in mind that the experimental error appears in the third decimal place, it is found that the values of Δs of four alkali metal chlorides shown in the figure almost coincide with each other. At a concentration of 1.0 mol kg^{-1}, the Δs values decrease by 0.01 mJ m^{-2} K^{-1}. These observations are very close to the predictions obtained from a linear regression line of the $d\gamma/dm$–temperature relation shown in Table 7.1. This variation in Δs for

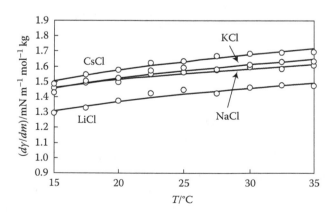

FIGURE 7.3 Variations of $d\gamma/dm$ of alkali metal chloride solutions with temperature.

TABLE 7.1

$[d(d\gamma/dm)/dT]/\mu JK^{-1}\ m^{-2}\ mol\ kg^{-1}$ Values Obtained from Figure 7.3 under the Assumption that the $d\gamma/dm$ Is a Linear Function of Temperature

	F	Cl	Br	I	NO_3	H_2PO_4	SO_4	HPO_4
Li	—	8.9	—	—	8.3	—	7.8	—
Na	4.0	7.3	5.5	9.8	11	12	12	18
K	—	9.1	—	—	13	6.5	13	13
Cs	—	10	—	—	—	—	25	—
NH_4	—	6.2	7.1	5.7	9.2	—	5.4	—
Mg	—	12	—	—	—	—	10	—
Ca	—	12	10	12	—	—	—	—
Cu	—	—	—	—	—	—	1	—
Ni	—	—	—	—	—	—	2	—

These values describe the difference in Δs between pure water and a 1 mol kg^{-1} solution that is $-(d\Delta s/dm)_{T,p}$.

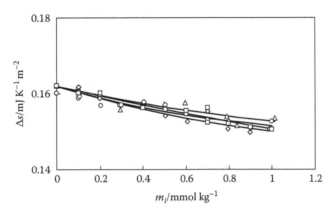

FIGURE 7.4 Entropy of surface formation for the aqueous solution of lithium (circle), sodium (triangle), potassium (square), and cesium (diamond) chlorides.

the simple electrolyte is considerably smaller than that encountered for the Δs of surface active substances shown in Section 4.4, but more significant than that observed for sucrose. The Δs values of the electrolyte solutions, smaller than Δs of pure water, indicate that the adsorption of ions into the surface region accompanies a negative contribution in the Δs. In other words, the partial molar entropies of ions in the surface region will be smaller than those in the bulk phase. This result follows from the fact that according to the surface potential measurements, there is an electrical double layer formed by anions and cations (Appendix A.2). The Δs values of the sucrose solutions are good proof that the entropy change corresponds to the formation of the electrical double layer.

7.2.2 SODIUM HALIDES

These systems are typical electrolyte solutions whose $d\gamma/dm$ depends on the magnitude of hydration or image force. The $d\gamma/dm$ of these salts increases with increasing temperature almost parallel to each other (Figure 7.5), and the lines connecting the plots are slightly concave downward. The magnitude of $d\gamma/dm$ decreases in the order NaF > NaCl > NaBr > NaI. The larger value in $d\gamma/dm$ suggests the larger magnitude of hydration of ions in the surface region; the aqueous solution of sodium fluoride has a large negative surface deficiency parallel with its strong hydration force and probably has a smaller surface density of ions. The magnitude of Δs is presumed to depend not only on the interaction between anions and cations but also on the surface densities of ions. The slopes of the linear regression lines of $d\gamma/dm$–T curves, which show the $-(d\Delta s/dm)_{T,p}$ values, are shown in Table 7.1. It is seen that $-(d\Delta s/dm)_{T,p}$ of sodium fluoride is small while that of sodium iodide has a much larger value. Again the Δs evaluated from the slopes of linear least-square lines of γ–T curves of sodium halide solutions are plotted against concentration in Figure 7.6. We see that Δs decreases almost linearly, and variation of the Δs can be well described by both methods.

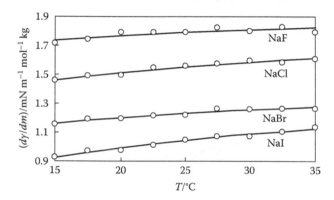

FIGURE 7.5 Variations of $d\gamma/dm$ of sodium halide solutions with temperature.

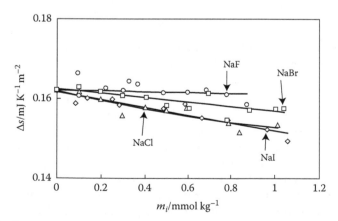

FIGURE 7.6 Entropy of surface formation for the aqueous solution of sodium fluoride (circle), chloride (triangle), bromide (square), and iodide (diamond).

7.2.3 SODIUM CHLORIDE AT HEXANE/WATER INTERFACE

Further consideration of the variation of the entropy change will be helpful in view of the general behavior of 1:1 electrolytes. Ikeda et al. (1992) have determined the entropy change of the adsorption of NaCl at a hexane/water interface using the experimental γ–T relations between 10°C and 40°C and at concentrations up to about 1 mol kg^{-1}. The curves are slightly concave to the temperature axis so that a linear regression line cannot be used to evaluate Δs. We have recalculated their data on the Δs values using their surface tension data between 15°C and 35°C and the concentrations up to 1 mol kg^{-1}. The Δs values are plotted in Figure 7.7 together with those of NaCl at the air/water interface. Since the values of Δs when the salt concentration is zero suggest the entropy of surface formation of air/water and hexane/water, respectively, the two different curves are drawn in the figure. The lower Δs–m curve for the hexane/solution interface first decreases steeply with concentration and thereafter decreases slightly and steadily, while the curve for the air/water interface decreases almost linearly. The value of Δs for the air/solution interface is almost twice that for the hexane/solution interface.

The $d\gamma/dm$ data for both the air/solution and the hexane/solution interface are compared in Figure 7.8. The solid line and dotted lines represent the graphs of the air/solution and the hexane/solution interface, respectively. The magnitude of $d\gamma/dm$ of these two graphs and their dependence upon temperature are very striking. The influence of an oil phase and that of an air phase upon $d\gamma/dm$ are strictly the same. Assuming that the linear least-squares method can be used, the evaluated $-(d\Delta s/dm)_{T,p}$ has roughly the same values 7.3 and 9.9 for the air/solution and the hexane/solution interface, respectively. In the surface region at the hexane/solution interface, we believe that there is an electrical double layer, which is exactly the same with that formed at the air/water interface.

Ikeda et al. (1992) also have measured the surface tension of the aqueous sodium chloride solutions as a function of pressure. With the aid of their reported data, fitted to quadratic empirical equations, we have estimated the surface tension–concentration

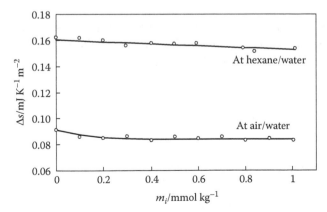

FIGURE 7.7 Comparison of the entropy of surface formation–concentration curves of the aqueous solutions of sodium chloride between air/water and hexane/water interfaces.

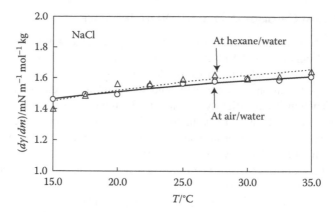

FIGURE 7.8 Comparison of $d\gamma/dm$–temperature curves of the aqueous solutions of sodium chloride between air/water (circles) and hexane/water (triangles) interfaces.

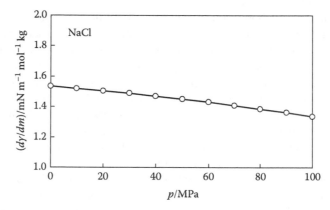

FIGURE 7.9 Variations of $d\gamma/dm$ of sodium halide solutions with pressure at the hexane/ water interface.

relations at the given pressures. The slope of the linear regression lines at the fixed pressure is plotted against pressure in Figure 7.9. For the surface tension–pressure variation, the reciprocity relation is given by

$$\left[\frac{d}{dp}\left(\frac{d\gamma}{dm}\right)_{T,p}\right]_{T,m} = \left[\frac{d}{dm}\left(\frac{d\gamma}{dp}\right)_{T,m}\right]_{T,p}. \tag{7.5}$$

The right side of this equation can be rewritten as

$$\left[\frac{d}{dp}\left(\frac{d\gamma}{dm}\right)_{T,p}\right]_{T,m} = \left(\frac{d\Delta v}{dm}\right)_{T,p}. \tag{7.6}$$

Thus, we can calculate the values of decrease in Δv with concentration at a given pressure from the slope of the $d\gamma/dm$–p graph. The $d\gamma/dm$ values decrease with pressure, and the curve is slightly concave to the pressure axis. Using the Δv value 0.0345 mm³ m⁻² of a pure hexane/water interface at 0.1 MPa, Δv decreases to 0.0332 at the concentration of 1 mol kg⁻¹ with variation by 0.0013. This value corresponds to the decrease in Δs; the partial molar entropy and volume of sodium chloride in the electrical double layer in the surface region are smaller than those in the bulk solution.

7.2.4 POLYATOMIC ANIONS AND CATIONS

The results of the electrolytes of the polyatomic anions are somewhat complex. The $d\gamma/dm$ values of nitrates and dihydrogen phosphates are plotted against temperature in Figure 7.10. Lithium, sodium, and potassium nitrates show almost the same $d\gamma/dm$ values and shape of the curve, but the calculated slopes increase in the order Li < Na < K (Table 7.1), which is in the order of $-\Gamma_i^H$ and is the reversal in the order of Γ_i^I.

In Figure 7.11, the $d\gamma/dm$ values of ammonium chloride, bromide, nitrate, and iodide are plotted against temperature. It is seen that the magnitudes of $d\gamma/dm$ and $-\Gamma_i^H/m_i$ do not explain the values of $[d(d\gamma/dm)/dT]$ shown in Table 7.1. They are probably complicated based on the relation between electrostatic interaction and chemical structures in the surface region.

Generally, it is tacitly understood that $d\gamma/dm$ is dependent upon the magnitude of hydration or interionic interaction with image force. However, on account of the earlier considerations, we will assume that molecular interactions between anions and cations in the surface region are important, which account for the formation of an electrical double layer in the surface region.

7.2.5 AQUEOUS SOLUTIONS OF 2:1 ELECTROLYTES

The earlier considerations demonstrate the entropic contribution and the significance of interionic interactions in the surface region. This will become more

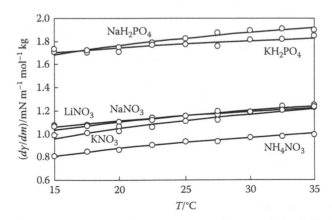

FIGURE 7.10 Comparison of $d\gamma/dm$–temperature curves of the aqueous solutions of alkali metal nitrate and dihydrogen phosphates.

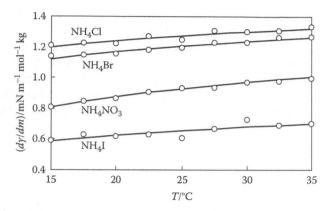

FIGURE 7.11   Comparison of $d\gamma/dm$–temperature curves of ammonium halide solutions.

apparent to us as we consider the $d\gamma/dm$–temperature graphs for 2:1 electrolytes. By analogy with the case of sodium halides, we expect that the magnitude of $[d(d\gamma/dm)/dT]$ will depend in part on the hydration of anions. In fact, the magnitude of $d\gamma/dm$ increases with the hydration of anions in the order $CaI_2 <$ $CaBr_2 < CaCl_2$. In Figure 7.12, the $d\gamma/dm$ values of these three salts are plotted against temperature. It is obvious that the slopes of these three calcium halides and magnesium chloride are nearly equal to each other. Linear regression lines have almost the same values for $MgCl_2$, $CaCl_2$, $CaBr_2$, and CaI_2, respectively, as shown in Table 7.1. It is undoubtedly true that this behavior is distinct from that of 1:1 alkali metal halides. This observation indicates that the double layer formations of these 2:1 electrolytes accompany remarkable but similar changes in the entropy. Since this entropy change as a result of the formation of the double layer will be caused by the force between divalent cations and univalent anions

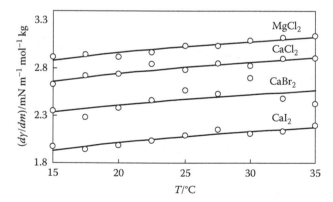

FIGURE 7.12   Comparison of $d\gamma/dm$–temperature curves of 2:1 electrolytes. From top to bottom, magnesium chloride, calcium chloride, bromide, and iodide, respectively.

in the layer, it is obvious that interionic interaction in the surface region depends upon the valence type of the 2:1 electrolytes.

7.2.6 AQUEOUS SOLUTIONS OF 1:2 ELECTROLYTES

In contrast to cations, a limited number of divalent anions are available. Alkali metal sulfates are typical divalent anions, but a satisfactory explanation for the shape of the curve connecting the plots of $d\gamma/dm-\Delta H_h$ relations is not available. In Figure 7.13, we have plotted $d\gamma/dm$ against temperature for lithium, sodium, potassium, cesium, and ammonium sulfate solutions. These curves are clearly concave downward, but the slopes of these curves are calculated from the linear regression lines and shown in Table 7.1. The magnitudes of $d\gamma/dm$ are in the order $Cs_2SO_4 > Li_2SO_4 > K_2SO_4 > Na_2SO_4$. However, the magnitude of the $[d(d\gamma/dm)/dT]$ decreases in the order $Cs_2SO_4 > K_2SO_4 > Na_2SO_4 > Li_2SO_4$. Again, there is no good correlation between the entropy change and the magnitude of the surface deficiency. Let us compare the two systems: the alkali metal sulfates and chlorides. It will be observed that the variations of Δs with concentration and cation species are sufficiently pronounced in the former system, so it seems possible that interionic interactions in the surface region of 1:2 electrolytes are distinct from those of 1:1-type alkali metal chlorides. The difference between CsCl and Cs_2SO_4 is very marked.

7.2.7 AQUEOUS SOLUTIONS OF 2:2 ELECTROLYTES

We have only three examples for 2:2 electrolytes: $MgSO_4$, $CuSO_4$, and $NiSO_4$. The $d\gamma/dm-T$ curves of these three electrolytes are shown in Figure 7.14. The behavior of $MgSO_4$ is similar to those of alkali metal chlorides. However, no satisfactory explanation of the validity of the data for $CuSO_4$ and $NiSO_4$ is available because of the limited experimental data.

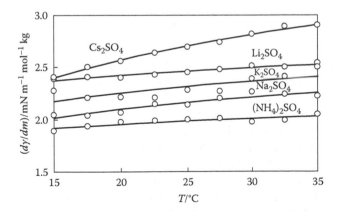

FIGURE 7.13 Comparison of $d\gamma/dm$–temperature curves of lithium, sodium, potassium, cesium, and ammonium sulfate solutions.

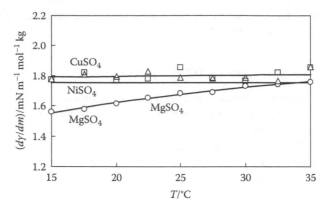

FIGURE 7.14 Comparison of $d\gamma/dm$–temperature curves of magnesium, copper, and nickel sulfate solutions.

7.3 HELMHOLTZ FREE ENERGY, ENTHALPY, AND ENERGY OF THE ADSORPTION OF ELECTROLYTES

When the system is in equilibrium, the Helmholtz free energy of adsorption is given by Equation 2.120:

$$\Delta f = \gamma - p\Delta v. \tag{2.120}$$

The magnitudes of the surface tension and the $p\Delta v$ term are available in Table 3.3. Generally, the values of $p\Delta v$ of the condensed phase are negligible in comparison with the surface tension, so the work content of the surface formation or adsorption is determined by the value of surface tension. Since heat content Δh is equal to $T\Delta s$ when the system is in equilibrium, the internal energy of adsorption is calculated by

$$\Delta u = \Delta f + \Delta h. \tag{7.7}$$

Comparison of these three terms of the adsorption of the aqueous sodium chloride solution is shown in Figure 4.26. The Δf, Δh, and Δu for a pure air/water interface have large values of 72, 48, and 120 mJ m^{-2} at 25°C, respectively, and are almost constant over the whole concentration range. It is difficult to illustrate their small variations in the figure. Thus, let us use the difference $\Delta y - \Delta y_0$ of these thermodynamic quantities of adsorption, where Δy_0 represents those of a pure air/water interface.

First, let us consider the adsorption of sodium iodide at the air/water interface whose hydration force is weak. The $(\Delta f - \Delta f_0)$, $(\Delta h - \Delta h_0)$, and $(\Delta u - \Delta u_0)$ of NaI solutions are plotted against concentration in Figure 7.15. It is probably true that the adsorption of ions increases almost linearly with the increasing bulk concentration; consequently, the figure illustrates the variation of these thermodynamic quantities along with the increasing concentration of ions in the surface region. The $(\Delta f - \Delta f_0)$ value of the NaI solution is positive, and the partial molar Helmholtz free energy has

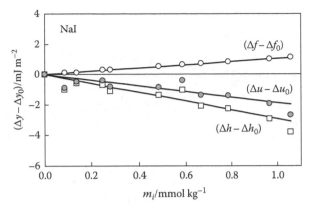

FIGURE 7.15 Deviations of energy, heat content, and work content of surface formation of sodium iodide solutions from those of a pure air/water interface.

a positive contribution to the system. However, the plots of the heat content and internal energy yield two almost straight lines with negative slopes. We can point out that the adsorption of ions accompanies an exothermic process and that the double layer formation in the surface region is also an exothermic process. The positive contribution of the work contents is compensated by this heat content and then the adsorption of ions in the surface region is accompanied by the decreases in the internal energy. The partial molar energy of ions in the surface region will be smaller than that in the bulk aqueous solution when the system is in equilibrium at the given temperature, pressure, and concentration.

An exceptional case of a different behavior of $(\Delta u - \Delta u_0)$ is observed for sodium fluoride solutions. $(\Delta f - \Delta f_0)$, $(\Delta h - \Delta h_0)$, and $(\Delta u - \Delta u_0)$ of NaF solutions are compared in Figure 7.16. In this case, in contrast to other electrolytes, $(\Delta h - \Delta h_0)$ does not have sufficient negative values, and as a result, $(\Delta u - \Delta u_0)$ has positive values. The partial molar energy of ions in the surface region will be larger than that in the bulk aqueous solution. The fluorine anion is classified as a typical structure maker anion, and the adsorption will be insignificant. There is another 1:1 electrolyte that

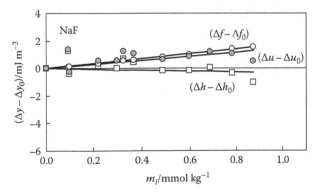

FIGURE 7.16 Deviations of energy, heat content, and work content of surface formation of sodium fluoride solutions from those of a pure air/water interface.

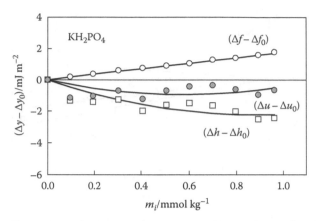

FIGURE 7.17   Comparison of energy, work contents, and heat contents of surface formation of KH_2PO_4.

serves as a structure maker anion, KH_2PO_4, whose $d\gamma/dm$ is significant. However, $(\Delta u - \Delta u_0)$ does not have positive values because $(\Delta h - \Delta h_0)$ has relatively significant negative values (Figure 7.17). The behavior of NaI illustrated in Figure 7.15 is a typical example of heat contents, work contents, and energy of the adsorption of simple electrolytes. It is to be noted that the variations of the thermodynamic quantities are too small to be found by surface tension measurements. Each step of calculations will accompany errors, and the individuality of ions becomes insignificant. When we try to describe the $(\Delta y - \Delta y_0)$ of single electrolyte solutions as shown in these figures, we find that there is no significant individual relationship between $(\Delta u - \Delta u_0)$, $(\Delta h - \Delta h_0)$, and $(\Delta f - \Delta f_0)$ except for NaF.

REFERENCES

Frumkin, A. Z. 1924. Phasengrenzkräfte und adsorption an der trennungsfläche luft Lösung anorganisheher electrolyte. *Z. Phys. Chem.* 109: 34–48.
Harkins, W. D. and E. C. Gilbert. 1926. The structure of films of water on salt solutions. II. The surface tension of calcium chloride solutions at 25°. *J. Am. Chem. Soc.* 48: 604–607.
Harkins, W. D. and H. M. McLaughlin. 1925. The structure of films of water on salt solutions I. Surface tension and adsorption for aqueous solutions of sodium chloride. *J. Am. Chem. Soc.* 47: 2083–2089.
Ikeda, N., M. Aratono, and K. Motomura. 1992. Thermodynamic study on the adsorption of sodium chloride at the water/hexane interface. *J. Colloid Interface Sci.* 149: 208–215.
Jarvis, N. L. and M. A. Scheiman. 1968. Surface potentials of aqueous electrolyte solutions. *J. Phys. Chem.* 72: 74–78.
Langmuir, I. 1917. Fundamental properties of solid and liquid II. *J. Am. Chem. Soc.* 39: 1848–1906.
Randles, J. E. B. 1957. Ionic hydration and the surface potential of aqueous electrolytes. *Discuss. Faraday Soc.* 24: 194–199.

8 Adsorption from Mixed Electrolyte Solutions

In the preceding chapters, the surface tension and the related thermodynamic quantities of two-component two-phase systems were discussed in detail. The thermodynamic quantities derived from the surface tension measurements suggest that the variation in the surface tension depends not only on the number of moles of ions in the solution but also on the interactions between constituent species in the surface region. There are meaningful distinctions in the character of interionic interactions between different valence types of ions. The individuality of interionic interactions between anions and cations observed for 1:1, 1:2, 2:1, and 2:2 electrolytes persists in the surface region, although the thermodynamic quantities of salts are less accurate than those for the surfactants. The point of view resulting from the earlier consideration suggests that there may be a lateral interaction between anions or cations in the surface region. Consideration of the aqueous solution of a mixture of electrolytes may be helpful, considering the lateral interaction of ions at the surface.

In most experimental works on the electrolytes in the surface region, the dominant variable is concentration, and our main attention has been concentrated on the increment of the surface tension produced by a single electrolyte. The temperature effect on the variation of surface tension of an aqueous salt solution has not attracted much investigation so far. For mixtures of inorganic salts, it is very hard to find literature that considers the temperature effect on surface tension. In a previous chapter, we demonstrated the significance of the temperature effect on surface tension that enables the evaluation of thermodynamic quantities such as entropy and energy of surface formation. By estimation of such a quantity defined as the thermodynamic quantity of adsorption, it is possible to obtain information of the surface region by means of which the interactions between components may be discussed. In this chapter, we intend to consider the thermodynamic quantities of surface formation of binary salt mixtures.

Ikeda (1977) has demonstrated the Gibbs adsorption isotherms for a mixture of an ionic surfactant and a simple univalent salt with a common ionic species, and Ikeda and Okuda (1988) developed the relations applicable to the multivalent salts. The thermodynamic theory of electrolyte solutions has been extended to the treatment of generalized mixtures applicable for any valence types (Motomura et al. 1990, Motomura and Aratono 1993). Let us consider a system in which the aqueous solution of two simple electrolytes, denoted (1) and (2), is in equilibrium with an air phase, provided that the mutual solubility of the aqueous solution and air is negligible. The surface tension of the solution will be given by

$$d\gamma = -s^H dT + v^H dp - \Gamma_{1+}^H d\mu_{1+} - \Gamma_{1-}^H d\mu_{1-} - \Gamma_{2+}^H d\mu_{2+} - \Gamma_{2-}^H d\mu_{2-}. \qquad (8.1)$$

When two bulk phases are practically immiscible and salts are insoluble in the air phase, the chemical potential of ion ik is given by

$$d\mu_{ik} = -s_{ik}^w dT + v_{ik}^w dp + \left(\frac{\partial \mu_{ik}}{\partial m_1}\right)_{T,p,m_2} dm_1 + \left(\frac{\partial \mu_{ik}}{\partial m_2}\right)_{T,p,m_1} dm_2 \quad i = 1,2, \quad (8.2)$$

where
   subscript i represents the electrolytes 1 and 2
   k represents the anion and cation
   subscripts 1+, 1−, 2+, and 2− represent cation and anion of components (1) and
      (2), respectively

For simplicity, our derivation will be restricted to the case of dilute ideal solutions where chemical potential of ion ik is represented by

$$\mu_{ik} = \mu_{ik}^0(T, p) + RT \ln m_{ik}. \quad (8.3)$$

According to this and the relation, $m_{ik} = \nu_{ik} m_i$, it holds that

$$\left(\frac{\partial \mu_{ik}}{\partial m_i}\right)_{T,p,m_j} = \left(\frac{\partial \mu_{ik}}{\partial m_{ik}}\right)\left(\frac{\partial m_{ik}}{\partial m_i}\right) = \frac{RT}{m_{ik}}\nu_{ik} = \frac{RT}{m_i}, \quad (8.4)$$

$$\left(\frac{\partial \mu_{ik}}{\partial m_j}\right)_{T,p,m_i} = 0. \quad (8.5)$$

We can write Equation 8.2 as

$$d\mu_{ik} = -\overline{s}_{ik}^w dT + \overline{v}_{ik}^w dp + \frac{RT}{m_i} dm_i. \quad (8.6)$$

When we consider the binary mixtures of salts with respect to common ions, mixtures with and without common ions should be considered separately. Further, when we proceed with the experimental work, there are two commonly used choice of variables: (1) m_1 and m_2 and (2) total concentration and composition of the second component. These four cases are considered separately in the following sections.

8.1   SURFACE TENSION OF BINARY MIXTURES AS A FUNCTION OF m_1 AND m_2

8.1.1   Mixtures with No Common Ion

First, let us illustrate the variation of surface tension of an electrolyte solution with no common ions as a function of T, p, m_1, and m_2. Substituting Equation 8.6 into Equation 8.1, the total differential of the surface tension is

$$d\gamma = -\Delta s dT + \Delta v dp - \frac{RT}{m_1}(\nu_{1+} + \nu_{1-})\Gamma_1^H dm_1 - \frac{RT}{m_2}(\nu_{2+} + \nu_{2-})\Gamma_2^H dm_2. \quad (8.7)$$

Here, Δs and Δv are

$$\Delta s = -\left(s^H - \Gamma_1^H \overline{s}_1 - \Gamma_2^H \overline{s}_2\right)dT \qquad (8.8)$$

and

$$\Delta v = \left(v^H - \Gamma_1^H \overline{v}_1 - \Gamma_2^H \overline{v}_2\right)dT, \qquad (8.9)$$

where we used the following relations:

$$\Gamma_{ik}^H = v_{ik}\Gamma_i^H \qquad (8.10)$$

and

$$\overline{y}_i = v_{i+}\overline{y}_{i+} + v_{i-}\overline{y}_{i-}. \qquad (8.11)$$

Equation 8.7 is useful in the discussion of the effect of salt on an ionic solute when there is no specific interaction between the salt and the solute. An effect of the solvent properties on the behavior of the solute can be developed. However, this relation cannot be useful when both have common ions or specific interactions.

8.1.2 MIXTURES WITH A COMMON ION

Next let us illustrate the surface tension of a mixed electrolyte solution with a common anion. We may write the total differential of the surface tension in the form

$$d\gamma = -s^H dT + v^H dp - \Gamma_{1+}^H d\mu_{1+} - \Gamma_{2+}^H d\mu_{2+} - \Gamma_{c-}^H d\mu_{c-}, \qquad (8.12)$$

where subscript c^- refers to the common anion. The differential of the chemical potential of cations is given by

$$d\mu_{i+} = -\overline{s}_{i+}dT + \overline{v}_{i+}dp + \frac{RT}{m_i}dm_i. \qquad (8.13)$$

Equation 8.4 for anion c^- can be written in the form

$$\left(\frac{\partial \mu_{c-}}{\partial m_i}\right)_{T,p,m_j} = \left(\frac{\partial \mu_{c-}}{\partial m_{c-}}\right)\left(\frac{\partial m_{c-}}{\partial m_i}\right)_{m_j} = \frac{RT}{m_{c-}}v_{i-}, \qquad (8.14)$$

where

$$m_{c-} = v_{1-}m_1 + v_{2-}m_2. \qquad (8.15)$$

The total differential of the common anion is given by

$$d\mu_{c-} = -s_{c-}dT + v_{c-}dp + \frac{RT}{m_{c-}}v_{1-}dm_1 + \frac{RT}{m_{c-}}v_{2-}dm_2. \qquad (8.16)$$

Substitution of Equations 8.13 and 8.16 into Equation 8.12 and rearrangement yield

$$dy = -\Delta s dT + \Delta v dp - RT\left(\frac{v_{1+}\Gamma_1^H}{m_1} + \frac{v_{1-}\Gamma_{c-}^H}{m_{c-}}\right)dm_1 - RT\left(\frac{v_{2+}\Gamma_2^H}{m_2} + \frac{v_{2-}\Gamma_{c-}^H}{m_{c-}}\right)dm_2.$$

(8.17)

For a binary mixture with common cation, we have

$$dy = -RT\left(\frac{v_{1+}\Gamma_{c+}^H}{m_{c+}} + \frac{v_{1-}\Gamma_1^H}{m_1}\right)dm_1 - RT\left(\frac{v_{2+}\Gamma_{c+}^H}{m_{c+}} + \frac{v_{2-}\Gamma_2^H}{m_2}\right)dm_2.$$

(8.18)

From an experimental point of view, it is desirable to reduce the complexity of this equation. The bracket of the third and fourth terms contains two unknowns Γ_1^H and Γ_2^H, respectively, and this will lead to the complexity of the practical experimental procedures.

8.2　SURFACE TENSION OF BINARY MIXTURES AS A FUNCTION OF CONCENTRATION AND COMPOSITION

In studying the miscibility of ionic surfactants in binary mixtures, Motomura et al. (1992) illustrated the usefulness of the variables, total number of moles of ions, m_{ti}, and composition X_2. Relations between m_{ti} and X_2, and an ordinary expression m_i, are

$$m_{ti} = m_{1+} + m_{1-} + m_{2+} + m_{2-} = v_1 m_1 + v_2 m_2$$

(8.19)

and

$$X_2 = \frac{v_2 m_2}{v_1 m_1 + v_2 m_2} = \frac{v_2 m_2}{m_{ti}},$$

(8.20)

where

$$v_i = v_{i+} + v_{i-} \quad i = 1, 2.$$

(8.21)

Differentiation of these equations yields

$$dm_{ti} = v_1 dm_1 + v_2 dm_2$$

(8.22)

and

$$dX_2 = \frac{-v_1 v_2 m_2 dm_1 + v_1 v_2 m_1 dm_2}{[v_1 m_1 + v_2 m_2]^2}.$$

(8.23)

These equations can be solved for variables m_1 and m_2 as

$$dm_1 = \frac{X_1}{v_1} dm_{ti} - \frac{m}{v_1} dX_2 \qquad (8.24)$$

and

$$dm_2 = \frac{X_2}{v_2} dm_{ti} + \frac{m}{v_2} dX_2. \qquad (8.25)$$

Here, the surface excess density of ions and composition in the surface region is defined as follows:

$$\Gamma_{ti}^H = \Gamma_{1+}^H + \Gamma_{1-}^H + \Gamma_{2+}^H + \Gamma_{2-}^H = v_1 \Gamma_1^H + v_2 \Gamma_2^H = \Gamma_{ti}^H X_1^H + \Gamma_{ti}^H X_2^H, \qquad (8.26)$$

$$X_2^H = \frac{v_2 \Gamma_2^H}{v_1 \Gamma_1^H + v_2 \Gamma_2^H} = \frac{v_2 \Gamma_2^H}{\Gamma_{ti}^H}. \qquad (8.27)$$

8.2.1 BINARY MIXTURES WITH NO COMMON ION

Substitution of Equations 8.24 and 8.25 into Equation 8.7 and rearrangement yield the expression for the binary mixtures with no common ion:

$$d\gamma = -\Delta s dT + \Delta v dp - \frac{RT}{m_{ti}} \Gamma_{ti}^H dm_{ti} - RT\Gamma_{ti}^H \left[\frac{X_2^H - X_2}{X_1 X_2} \right] dX_2. \qquad (8.28)$$

8.2.2 MIXTURES WITH A COMMON ANION

Substitution of Equations 8.24 and 8.25 into Equation 8.17 and rearrangement yield the expression for the binary mixtures. The surface tension at the given temperature and pressure is given by lengthy rearrangement as

$$(d\gamma)_{T,p} = -RT \frac{\Gamma^H}{m} dm + RT \frac{\Gamma^H}{X_1 X_2} \left(X_2 - X_2^H \right) \left[1 - \frac{v_1 v_{2-}}{v_2 v_{1-} X_1 + v_1 v_{2-} X_2} \right] dX_2. \qquad (8.29)$$

For the mixtures of 1:1 electrolytes with a common ion, where $v_1 = 2$, $v_2 = 2$, $v_{1+} = 1$, $v_{1-} = 1$, $v_{2+} = 1$, $v_{2-} = 1$, Equation 8.29 reduces to

$$(d\gamma)_{T,p} = -\frac{RT}{m_{ti}} \Gamma_{ti}^H dm_{ti} - \frac{RT\Gamma_{ti}^H}{2} \left[\frac{X_2^H - X_2}{X_1 X_2} \right] dX_2. \qquad (8.30)$$

It is to be noted that this equation can also be applicable for mixtures of 1:1 electrolytes with a common cation.

For a mixture of 1:1 and 2:1 electrolytes with a common anion where $\nu_1 = 2$, $\nu_2 = 3$, $\nu_{1+} = 1$, $\nu_{1-} = 1$, $\nu_{2+} = 1$, $\nu_{2-} = 2$, Equation 8.29 reduces to

$$(d\gamma)_{T,p} = -\frac{RT}{m_{ti}}\Gamma_{ti}^{H} - \frac{RT\Gamma_{ti}^{H}}{X_1 X_2}\left[\frac{\left(X_2^{H} - X_2\right)(1 + X_2)}{(3 + X_2)}\right](dX_2). \tag{8.31}$$

These relationships are more easily derived using the total differential of the chemical potentials of ions with respect to T, p, m_{ti}, and X_2.

8.3 MIXTURES OF ELECTROLYTES WITH COMMON CATION

8.3.1 SODIUM CHLORIDE–SODIUM IODIDE MIXTURES

First let us consider the solutions of mixtures of two 1:1 electrolytes with a common cation at constant composition (Matubayasi et al. 2001). The surface tension of equimolar mixtures of NaCl and NaI is plotted against concentration in Figure 8.1 in order to show that linear regression lines are applicable to these relations in a similar manner to a single electrolyte solution. Figure 8.2 compares the $d\gamma/dm_t$ values of NaCl–NaI mixtures with that of single salt solutions of NaCl and NaI, respectively. Here, m_t represents total molality that is the sum of m_1 and m_2. We have used the total number of moles of the electrolytes instead of the total number of ions in the solution. A thin dotted line drawn in the figure shows a value for an arithmetic mean of the $d\gamma/dm$ value of NaCl and NaI observed for single salt solutions. In the previous chapter, we have used the linear regression lines of the $d\gamma/dm$–T plots in the considerations of the entropy of adsorption of electrolyte at the surface, because the $d\gamma/dm$–T plots of single electrolyte solutions are almost straight without exception. The solid line connecting the plots for mixtures is a quadratic regression line. At lower temperatures, $d\gamma/dm_t$ of the anion mixture shows the characteristic behavior of simple 1:1 electrolytes and varies along the line of the arithmetic mean; however, it starts to deviate from the line around about 25°C and thereafter plots approach to the line of iodide ions with increasing temperature.

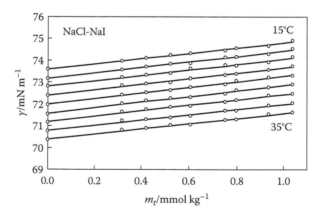

FIGURE 8.1 Surface tension–concentration curves for the aqueous solution of equimolar mixtures of sodium chloride and sodium iodide. Temperatures are 15.0°C, 17.5°C, 20.0°C, 22.5°C, 25.0°C, 27.5°C, 30.0°C, 32.5°C, and 35°C from top to bottom.

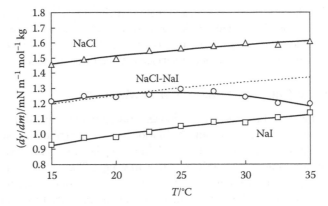

FIGURE 8.2 Comparison $d\gamma/dm$–temperature curve of equimolar mixture of sodium chloride and sodium iodide (circle) with that of sodium chloride (triangle) and that of sodium iodide (square).

This observation clearly produces a characteristic image for the mixtures. The negative slope of the curve suggests that Δs will increase with increasing temperature and that the partial molar entropy of the electrolyte has positive values. It is to be noted that less significant dependence of the activity coefficients of sodium halides on temperature is shown in the textbook of Harned and Owen (1958, p. 506).

The surface tension of the anion mixtures linearly decreases with increasing temperature. With the method of least squares, the values of Δs are obtained from the slopes of the γ–T plots and are plotted against concentration in Figure 8.3 together with those of NaCl and NaI solutions. The deviation from the almost linear variation of the single electrolyte solution is clearly seen. In a dilute solution, there is a rapid decrease in Δs until a minimum appears. After this, the Δs value rises over that of a pure water/air surface. This behavior may be a characteristic of the anion mixture of simple electrolyte solutions. The positive slope corresponds to negative values of $d(d\gamma/dm_t)/dT$. The negative slope of $d\Delta s/dm_t$ observed for a dilute solution suggests that the $d(d\gamma/dm_t)/dT$ must

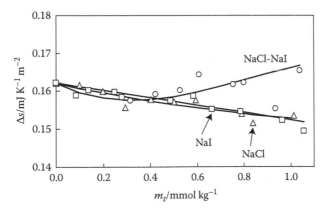

FIGURE 8.3 Entropy of surface formation for the aqueous solution of an equimolar mixture of sodium chloride and sodium iodide (circle) with that of sodium chloride (triangle) and that of sodium iodide (square).

be positive; in other words, the conclusion that the $d(d\gamma/dm_t)/dT$ of mixture tends to be negative does not hold for a dilute solution of mixtures. These considerations lead to the fact that the surface tension of the mixture is not strictly a linear function of m_t, but it requires precision far beyond that of our surface tension measurements.

8.3.2 Sodium Chloride–Sodium Sulfate Mixtures

We have also measured the surface tension as a function of temperature and concentration of equimolar mixtures of sodium chloride and sulfate (Matubayasi et al. 2001). Linear regression lines are well fitted to γ–T and γ–m_t data in the limited temperature and concentration range: 15°C–35°C and 0–1 mol kg^{-1}. The $d\gamma/dm_t$–T curves of the mixtures are illustrated in Figure 8.4. Again we used the total concentration of the electrolytes m_t instead of the total number of ions m_{ti} in order to illustrate the difference between single electrolyte and mixed electrolyte solutions. If we use the m_{ti} and evaluate the $d\gamma/dm_{ti}$, we will find two overlapping $d\gamma/dm_t$–T curves of NaCl and Na$_2$SO$_4$. The $d\gamma/dm_{ti}$ of the equimolar mixture has a value very close to them at 15°C and begins to separate downward widely with increasing temperature. Figure 8.4 shows that the $d\gamma/dm_t$–T curve of the anion mixtures also has a negative slope, and the shape confirms that there is a characteristic behavior of the aqueous solution of simple electrolyte mixtures in the Δs. The Δs–m_t curve of the mixture is illustrated in Figure 8.5. Comparison of the shape of this curve with that for NaCl–NaI mixtures leads to the conclusion that Δs–m_t plots of which the curve is concave upward with clear minimum represent a distinctive shape for the anion mixtures.

8.4 MIXTURES OF ELECTROLYTES WITH A COMMON ANION

Throughout our consideration of simple electrolytes, we observed that anions are of relative importance and significance in the surface region. However, it is of interest to consider the information obtainable on the behavior of cation mixtures in the surface region, since we observed distinctive behaviors of univalent and divalent cations in previous chapters, respectively.

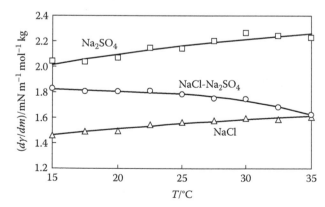

FIGURE 8.4 Variation of $d\gamma/dm$-temperature curve of equimolar mixture of sodium chloride and sodium sulfate (circle). Square and triangle represent those of pure sodium sulfate and sodium chloride solutions, respectively.

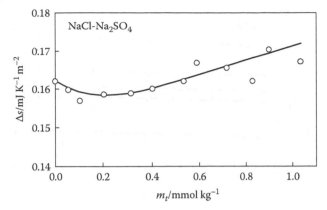

FIGURE 8.5 Entropy of surface formation for the aqueous solution of equimolar mixture of sodium chloride and sodium sulfate.

8.4.1 SODIUM CHLORIDE–MAGNESIUM CHLORIDE MIXTURES

The γ–m_t relations of the equimolar mixtures of $X_2 = 0.4996$ are shown in Figure 8.6. Within the observed ranges of m_t and T, linear empirical equations can be fitted to the plots, similar to those observed for single salt solutions, and the slope of the linear regression line increases with increasing T. The γ–T relationships of the mixtures also show a linear relation as for those observed for a single salt solution (Matubayasi et al. 1999).

Just like the behavior of anion mixtures of the simple electrolytes with a common cation, let us consider the graph of $d\gamma/dm_t$–T of the NaCl–MgCl$_2$ mixtures (Figure 8.7). There is a gradual increase in $d\gamma/dm_t$ of the mixtures with temperature. The variation of $d\gamma/dm_t$ with temperature for mixtures is not the behavior that may be expected to follow a pattern observed for the anion mixtures of the electrolyte solutions with a common cation. We repeatedly have observed a meaningful distinction in surface thermodynamic quantities between anions and cations. The anionic species

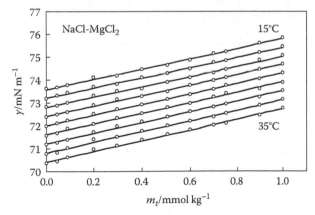

FIGURE 8.6 Surface tension–concentration curves for the aqueous solution of equimolar mixtures of sodium chloride and magnesium chloride. Temperatures are 15.0°C, 17.5°C, 20.0°C, 22.5°C, 25.0°C, 27.5°C, 30.0°C, 32.5°C, and 35°C from top to bottom.

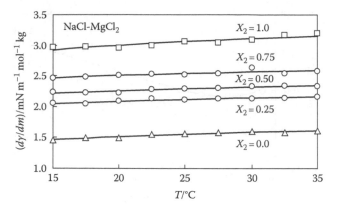

FIGURE 8.7 The $d\gamma/dm$–temperature relations of the mixtures of sodium chlorides (triangle) and magnesium chloride (square). The compositions of the mixtures are 0.0, 0.25, 0.50, 0.75, and 1.0 from bottom to top.

are of fundamental importance in determining magnitude of $d\gamma/dm$, Δs, and Γ_i^H/m_i, but difference in the cationic species is of less significance. Again, it is important to point out that cationic mixtures are of less fundamental importance in determining the shape of $d\gamma/dm_t$–T graphs.

We now turn our attention to the slope of the $d\gamma/dm_t$–T graphs. For anion mixtures, the $d\gamma/dm_t$ values are not a linear function of temperature, but a linear regression line can be applied for cation mixtures. We applied the method of least squares for the linear $d\gamma/dm_t$–T graphs, and the calculated slopes are plotted against the composition of $MgCl_2$, X_2, in Figure 8.8. The ordinate shows the magnitude of decrease in Δs, since $d(d\gamma/dm_t)/dT$ is equal to $-(d\Delta s/dm_t)$. The graph represents that Δs of the cation mixtures are always larger than those of single electrolyte solutions of NaCl and of $MgCl_2$ because of the mixing of cations in the surface region. It is probable that there are lateral interactions between cations in the surface region. This behavior is confirmed by the results observed for anion mixtures, although the $d\gamma/dm_t$–T curves of cation mixtures have positive slopes. Now let us plot the Δs values of mixtures derived directly

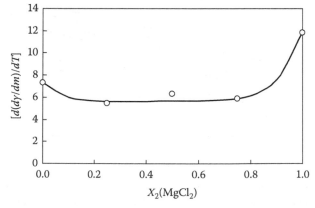

FIGURE 8.8 Variation of $[d(d\gamma/dm)/dT]$ that is $-(d\Delta s/dm)$ with the composition of $MgCl_2$.

from the γ–T curves by using the method of least squares (Figure 8.9). In order to avoid confusion in the figure, the data of single electrolyte solutions are omitted. The characteristic concentration dependence of Δs on concentration is clearly shown. As is seen in Figure 8.2 for anion mixtures, Δs first decreases with concentration, passes through the minimum, and increases with concentration. The appearance of a minimum in the curves is less clear than those of anion mixtures, but the result suggests that there are lateral interactions between cations in the surface region.

If there is no lateral interaction between adsorbed cations in the surface region, there is a linear relationship between Γ_2^H and X_2. However, the curve is concave downward and the surface deficiencies of the mixtures have a larger value than the line connecting the two electrolytes. The larger value can be visualized in terms of a diagram such as Figure 8.10, where X_2^H plotted against X_2 at γ is 73 mN m^{-1}. The surface deficiency of magnesium chloride is larger than that of sodium chloride, since X_2^H deviates upward from the dotted line $X = X_2$.

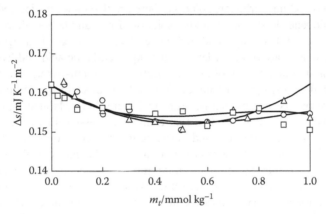

FIGURE 8.9 Variation of the entropy of surface formation with concentration. The compositions of the mixtures $X_2(MgCl_2)$ are 0.250 (circles), 0.500 (triangles), and 0.747 (squares), respectively.

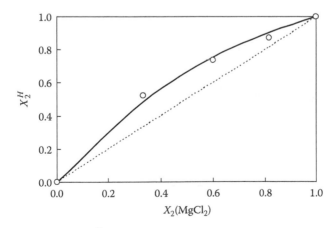

FIGURE 8.10 Variation of X_2^H with X_2.

8.5   SOLUTIONS WITH IMPURITY

In the earlier consideration of the behaviors of mixtures, we have considered the equimolar mixtures of simple electrolytes. We have another type of mixture. When salts of a weak acid or weak base are dissociated into water, hydrolysis occurs and forms mixtures of the electrolytes with a common anion or cation. In previous chapters, we have assumed that hydrolysis does not occur in the aqueous solutions of ammonium and sulfate salts. This assumption will be supported by the results that a positive linear regression line can be roughly fitted to the $d\gamma/dm$–T curves and that Δs shows a monotonic decrease function of concentration. In this section, let us now consider the mixtures of electrolytes resulting from the hydrolytic reaction whose $d\gamma/dm$–T curves will not be assumed as a solution of a single salt.

8.5.1   SODIUM CARBONATE SOLUTIONS

In an aqueous sodium carbonate solution, there are five ionic species and undissociated carbonic acid. In the 0.5 mol kg^{-1} solution, for example, the abundant species are sodium ions of 1 mol kg^{-1} and carbonate of 0.49 mol kg^{-1}; hydrogen carbonate and hydroxide ions are of the order of mmol kg^{-1}; carbonic acid and hydrogen ions are traces of 10^{-8} and 10^{-12} mol kg^{-1}, respectively. It seems likely that the solution can be treated as a single salt solution. The $d\gamma/dm$ values of the sodium carbonate solutions are plotted against temperature in Figure 8.11 (Matubayasi et al. 2001). The curve is concave downward so that a linear regression line cannot be drawn. The shape of the quadratic line used in the figure is similar to that used for the line in Figure 8.2, although the solution is not an equimolar mixture of univalent ions. The curve suggests that the solution is a mixture of divalent carbonate and univalent hydrogen carbonate anions. In Figure 8.12, we have plotted the Δs values derived directly from the γ–T curves against concentration. It is probable that the curve has a minimum characteristic for ionic mixtures in the surface region. This shape of the curve confirms that the behavior of $d\gamma/dm$ is shown in Figure 8.11. These observations suggest that even a small amount of co-ions or contamination could change the

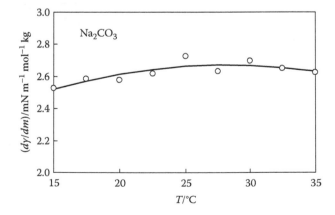

FIGURE 8.11   Variation of $d\gamma/dm$ of sodium carbonate solutions with temperature.

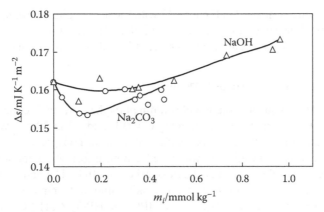

FIGURE 8.12 Variations of Δs values of sodium carbonate (circles) and sodium hydroxide (triangles) solutions with concentration.

$d\gamma/dm$ or γ values of aqueous salt solutions and that the values of these properties of mixtures depend upon the combination of mixing ions.

8.5.2 SODIUM HYDROXIDE SOLUTIONS

There is a description in a textbook for a laboratory course that sodium hydroxide must be free from contamination by sodium carbonate. In Figure 8.13, the $d\gamma/dm$ values of untreated sodium hydroxide solutions are plotted against temperature. It is observed that the $d\gamma/dm$ value is decreased more significantly with temperature than that of the sodium carbonate solution. The $[(d\gamma/dm)/\text{mN m}^{-1} \text{ mol}^{-1} \text{ kg}]$ values of untreated sodium hydroxide solutions are 1.86 compared to 1.98 for treated reagent. It is probable that the negative slope is accountable by the incorporation of carbon dioxide into the reagents or solutions during the experiments, although we have not measured the temperature effect for purified sodium hydroxide solutions.

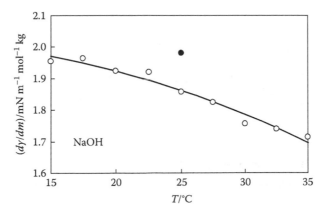

FIGURE 8.13 Variation of $d\gamma/dm$ of sodium hydroxide solutions with temperature. Open circles show the data for untreated sodium hydroxide, and the solid circle is the result of treated sodium hydroxide.

The Δs values are compared in Figure 8.12 with those of sodium carbonate solutions. These observations suggest that experimental data of our simple salt solutions presented in previous chapters are free from the effect of contaminations.

8.5.3 SODIUM PHOSPHATE SOLUTIONS

In the 0.5 mol kg^{-1} aqueous solution of sodium phosphate Na$_3$PO$_4$, concentrations of anions PO$_4^{3-}$, HPO$_4^{2-}$, and OH$^-$ are roughly 0.4, 0.1, and 0.1 mol kg^{-1}. We measured the surface tension of the aqueous solution of sodium phosphate as a function of temperature and concentration, because we have no information about mixed solutions like these complex mixtures (Matubayasi et al. 2011). The behavior of γ–T and γ–m relations is similar to those of a single salt solution. Comparison between the $d\gamma/dm$–T curves observed for Na$_3$PO$_4$, Na$_2$HPO$_4$, K$_2$HPO$_4$, NaH$_2$PO$_4$, and KH$_2$PO$_4$ solutions is made in Figure 8.14. The scattering of the plots for dibasic and tribasic phosphates is on account of their low solubility. In the consideration of the behavior of $d\gamma/dm$ for 1:1 and 1:2 electrolytes thus far, we obtained the following four characteristics:

1. The $d\gamma/dm$ value of 1:1 electrolytes is substantially independent of cation species.
2. The $d\gamma/dm$ value of 1:2 electrolytes is appreciably larger than that of 1:1 electrolytes.
3. The $d\gamma/dm$ value of 1:2 electrolytes depends on the cationic species.
4. The $d\gamma/dm$–T curves of 1:2 and 2:1 electrolytes have larger slopes than those of 1:1 electrolytes. It is obvious that we cannot expand these four criteria to include 1:3 electrolytes by using the data of tribasic phosphate Na$_3$PO$_4$. However, the curve shown in Figure 8.15 satisfies neither of the requirements for ionic mixtures:
   A. The $d\gamma/dm$–T curves of the ionic mixtures are concave downward.
   B. The Δs–m curves of the ionic mixtures are concave upward.

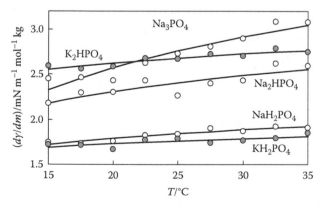

FIGURE 8.14 Comparison of $d\gamma/dm$–temperature curves of Na$_3$PO$_4$, Na$_2$HPO$_4$, K$_2$HPO$_4$, NaH$_2$PO$_4$, and KH$_2$PO$_4$ solutions.

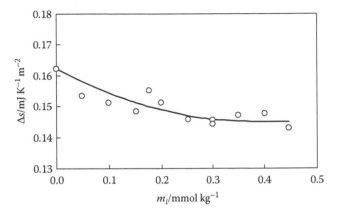

FIGURE 8.15 Variations of Δs of a sodium phosphate solution.

The slope $d(d\gamma/dm_i)/dT$ of the aqueous Na_3PO_4 solutions is larger than those of dibasic and monobasic phosphates, and the Δs value decreases more largely than those of monobasic and dibasic phosphates (Figure 8.15). Further, the $[(d\gamma/dm_{ti})/\text{mN m}^{-1} \text{ mol}^{-1} \text{ kg}]$ value of Na_3PO_4 solutions is 0.68, which is smaller than 0.8–0.9 of monobasic and dibasic phosphates. The behavior of Na_3PO_4 solutions is inexplicable.

REFERENCES

Harned, H. S. and B. B. Owen. 1958. *The Physical Chemistry of Electrolytic Solutions*, 3rd edn., American Chemical Society Monograph series. Reinhold Publishing Corporation, New York.

Ikeda, S. 1977. On the Gibbs adsorption equation for electrolyte solutions. *Bull. Chem. Soc. Jpn.* 50: 1403–1408.

Ikeda, S. and H. Okuda. 1988. Preferential adsorption of ions on aqueous surfaces of mixed solution of mono-monovalent and multi-multivalent strong electrolytes. *J. Colloid Interface Sci.* 121: 440–448.

Matubayasi, N., K. Tsunetomo, I. Sato, R. Akizuki, T. Morishita, and Y. Natuskari. 2001. Thermodynamic quantities of surface formation of aqueous electrolyte solutions IV. Sodium halides, anion mixtures, and sea water. *J. Colloid Sci.* 243: 444–456.

Matubayasi, N., R. Tsutsumi, H. Tachibana, and M. Kishimoto. 2011. Aqueous solution of phosphates, unpublished.

Matubayasi, N., S. Yamaguchi, K. Yamamoto, and H. Matsuo. 1999. Thermodynamic quantities of surface formation of aqueous electrolyte solutions II. Mixed aqueous solutions of NaCl and $MgCl_2$. *J. Colloid Sci.* 209: 403–407.

Motomura, K., N. Ando, H. Matsuki, and M. Aratono. 1990. Thermodynamic studies on adsorption at interfaces VII. Adsorption and micelle formation of binary surfactant mixtures. *J. Colloid Sci.* 139: 188–197.

Motomura, K. and M. Aratono. 1993. Miscibility in binary mixtures of surfactants. In *Mixed Surfactant Systems*, eds. K. Ogino and M. Abe, pp. 99–145. Marcel Dekker, Inc., New York.

Motomura, K., T. Kanda, K. Abe, N. Todoroki, N. Ikeda, and M. Aratono. 1992. Miscibility of dodecylammonium chloride and octylsulfinylethanol in the adsorbed film and micelle. *Colloids Surf.* 67: 53–59.

9 Aqueous Solutions of Zwitterionic Amino Acids

In the preceding chapters, the thermodynamic quantities necessary for the understanding of simple salt solutions have been considered. We will now attempt to demonstrate the thermodynamic properties of zwitterionic amino acids in the surface region. Until Bull and Breese (1974) reported careful measurement of the surface tension for a series of amino acids as a function of concentration, there was little knowledge of the properties of zwitterionic amino acids in the surface region, although the lateral interactions of zwitterionic surfactants are well studied. Pappenheimer et al. (1936) studied amino acid solutions, taking the large electric dipole moment of the zwitterion into account, by means of the drop volume method. They found that glycine, α-alanine, β-alanine, and β-amino butyric acids all behave like inorganic salt, and the surface tension of their solution is greater than that of water despite no net charge. Thereafter, this subject has been investigated by Belton (1939), who showed that $d\gamma/dm$ positively depends on the hydrocarbon chain length, and the variation becomes more marked in the presence of salt. In addition, Belton and Twidle (1940) reported the variation of $d\gamma/dm$ with pH for glycine and alanine solutions. In this chapter, we present a consideration of the difference between zwitterions and simple salts in the surface region. There are three cases for which an orientation of zwitterions at the surface must be sought. First, the difference in interactions with water dipoles between zwitterions and ions may be of interest. In Sections 9.1 and 9.2, comparison of the results obtained from the surface tension measurements for glycine solutions will be made with the simple electrolytes considered in the previous chapters. Next we will consider the lateral interactions of zwitterions aligned parallel to the surface. Amphipathic leucine molecules will be oriented approximately normal to the water surface, and the zwitterions will be aligned parallel to the surface; it will be of interest to consider the relation between two kinds of interactions, that is, forces between dipole–dipole and attractive forces between hydrophobic groups. Finally, the interaction between simple salt and the aligned dipoles at the surface will be considered in Section 9.4.

9.1 AQUEOUS SOLUTIONS OF GLYCINE

When glycine, amino acetic acid, and molecules are introduced into water, part of them are converted to $NH_3^+CH_2COOH$ and $NH_2CH_2COO^-$ by hydrolysis reaction, and the solution has a pH of 6.07 known as the isoelectric point of glycine because of the electrical neutrality condition (Appendix A.5). In Figure 9.1, the surface tension of glycine solutions are plotted against concentration up to 1 mol kg^{-1} and the temperature range of 15°C to 35°C at 2.5°C intervals. The surface tension increases

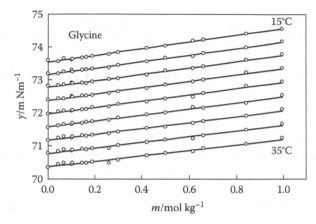

FIGURE 9.1 Variation of surface tension with concentration for the adsorption of glycine at air/water interface. Temperatures are 15.0°C, 17.5°C, 20.0°C, 22.5°C, 25.0°C, 27.5°C, 30.0°C, 32.5°C, and 35°C from top to bottom.

almost linearly in a similar manner to that of simple salt solutions. The $d\gamma/dm$–T relation obtained by the method of least squares is compared with that of ammonium nitrate in Figure 9.2. The [($d\gamma/dm$)/mN m^{-1} mol^{-1} kg] value of the solution at 25°C is 0.957—very close to that of NH_4NO_3, 0.927—but there is a significant distinction between them. The $d\gamma/dm$–T plots of the aqueous solution of NH_4NO_3 have positive slopes. Since the electrical double layer is formed as a result of the adsorption of ammonium nitrate ions, the partial molar entropy of NH_4NO_3 is smaller than that in the bulk solution. Clearly, in contrast to the NH_4NO_3 solution, the negative slope observed for glycine suggests that the partial molar entropy of glycine is larger than that in the bulk phase. Recalling the consideration of the curve with the anion mixtures of sodium sulfate and sodium chloride in the previous chapter, we can point out that the products of the hydrolysis or contamination contribute to this negative slope. This is probably true, since the $d\gamma/dm$–T curve with negative slope of the solution

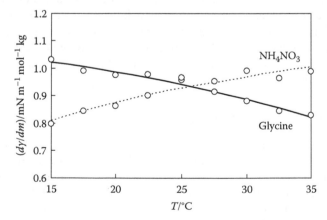

FIGURE 9.2 Comparison of $d\gamma/dm$–temperature curves of glycine (solid line) and ammonium nitrate (dotted line) solutions.

of a single solute may almost never be known for a single salt solution. As shown in Table A.5, the glycine solution contains a small amount of the anion $H_2NCH_2COO^-$ and the cation $NH_3^+CH_2COOH$. Thus we shall expect to find certain amount of these anionic and cationic glycines in the surface region with preferential of the zwitterionic one so that the surface tension behaves as a mixture of these glycines. It is interesting to compare the adsorption of zwitterionic glycine with ionic glycine, since the difference between these two types will show a clue to a possible explanation of how a dipole behaves at the surface.

The Δs values obtained directly from the γ–T curves are plotted against concentration in Figure 9.3. Δs first decreases to a minimum value with increasing concentration as if the system consists of a single salt solution, passes through a minimum, and increases with concentration according to the familiar behavior of the mixtures of two simple salts. This behavior leads to a consideration of a competitive adsorption between two components. At low concentrations, there is no marked influence of the second component in the entropy when the mixture is an ideal solution in the surface region because of the limited concentration of one or both components. A comparison of these considerations with those of the sodium chloride and sodium sulfate mixtures leads to the conclusion that a glycine dipole behaves as an ion pair of simple salts.

As an example of an amino acid, the Δs–m graph of alanine is also shown in Figure 9.3 for comparison with that of glycine, in which the curve has the same shape as that of glycine. It first decreases and immediately turns to increase with increasing concentration. However, it is to be noted that an increase in the hydrophobic group results in the flattening of the curve. We have not shown the graphs of the valine and leucine solutions, since their experimental concentration ranges are limited below 0.15 mol kg^{-1}, but the flattening of the curve is more clearly seen for these solutions.

It is to be noted that the adsorption of these surface active amino acids may be characterized by Δs, which has no deviation from that of pure water throughout the entire experimental concentration range. Lateral interactions between hydrophobic groups, which strongly influence the entropy of surfactant molecules, are not

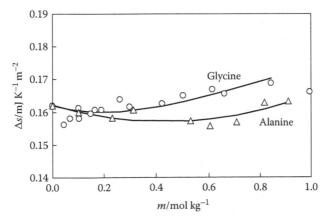

FIGURE 9.3 Variation of the entropy of surface formation with concentration. Glycine (circles); alanine (triangles).

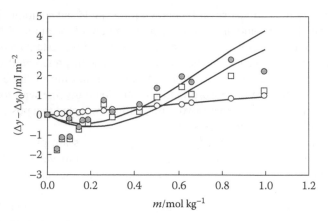

FIGURE 9.4 Variation of work content, heat content, and energy change for the adsorption of glycine with concentration. Open circles, $(\Delta f - \Delta f_0)$; squares, $(\Delta h - \Delta h_0)$; and filled circles, $(\Delta u - \Delta u_0)$.

observed because of lateral electrostatic interactions between adsorbed zwitterions of amino acids. These observations suggest that the Δs–m graph of glycine molecules is considered to be a characteristic of the simple electrolytes.

Now let us consider the work and heat contents of the adsorption of glycine. $(\Delta f - \Delta f_0)$, $(\Delta h - \Delta h_0)$, and $(\Delta u - \Delta u_0)$ are plotted against concentration in Figure 9.4. The work content $(\Delta f - \Delta f_0)$ increases linearly as a function of concentration. The heat content $(\Delta h - \Delta h_0)$ shows a minimum and increases steeply with increasing concentration. Since energy of adsorption is a sum of work contents and heat contents, a remarkable increase in $(\Delta u - \Delta u_0)$ is observed. Positive variation in $(\Delta u - \Delta u_0)$ suggests that the partial molar energy of glycine in the surface region has larger values than that in the bulk phase and that the adsorption is energetically unfavorable.

In the earlier consideration, we assumed that the glycine solutions are a mixture of zwitterionic and ionic glycine molecules. We have disregarded the possibility that the results observed for zwitterionic glycine are characteristics of dipoles in the surface region. It is to be noted that the negative slope shown in Figure 9.2 is remarkable, but so far, there is no proof that a simple dipole shows a positive change with the adsorption.

9.2 SURFACE TENSION OF HYDROCHLORIC ACID–GLYCINE AND GLYCINE–SODIUM HYDROXIDE SYSTEMS AT 25°C

Glycine is a typical dipolar molecule, and the aqueous solution behaves as though it is a mixture of simple salts. It is of interest to consider the information obtainable on a mixture of dipolar glycine and ionic glycine. The following abbreviations are used to denote these glycines; GH^+, $NH_3^+CH_2COOH$; $G^\pm$, $NH_3^+CH_2COO^-$, G^-, $NH_2CH_2COO^-$. In this section, we consider mixtures consisting of $GHCl$–$G^\pm$ and $G^\pm$–NaG, respectively.

The acid dissociation constant k_1 of the equilibrium

$$NH_3^+CH_2COOH \rightleftharpoons NH_3^+CH_2COO^- + H^+$$

is 4.47×10^{-3}. When excess HCl is added to the aqueous solution of zwitterionic glycine, most of the $G^\pm$ changes its form to glycine hydrochloride (GHCl). The second acidity constant k_2 for the equilibrium

$$NH_3^+CH_2COO^- \rightleftharpoons NH_2CH_2COO^- + H^+$$

also has a small value 1.66×10^{-10}. When excess NaOH is added to the aqueous solution of $G^\pm$, most of the $G^\pm$ changes to sodium glycinate, NaG. In a textbook, the pH is usually employed as an independent and dependent variable for the acid–base reactions, but it is to be noted that proton is not the major constituent that determines the properties of the solutions except in strongly acidic solutions. We will consider the number of moles of the major constituents instead of pH.

9.2.1 Zwitterionic Glycine ($G^\pm$) and Glycine Hydrochloride Mixtures

The system composed of glycine hydrochloride and zwitterionic glycine is a five-component system, that is, air, water, GH^+, Cl^-, and $G^\pm$. As independent variables, we use the total molality of the mixture defined by

$$m = m_{G^\pm} + m_{GHCl} \tag{9.1}$$

and the mole fraction of GHCl defined by

$$X = \frac{m_{GHCl}}{m_{GHCl} + m_{G^\pm}}. \tag{9.2}$$

At constant temperature and pressure, we can write the variation of γ of this system as

$$d\gamma = -\frac{RT\Gamma^H}{m}dm + \Gamma^H RT \left[\frac{X - X^H}{X(1-X)} \right] dX. \tag{9.3}$$

Here Γ^H and X^H are defined as

$$\Gamma^H = \Gamma^H_{GH^+} + \Gamma^H_{Cl^-} + \Gamma^H_{G^\pm} \tag{9.4}$$

and

$$X^H = \frac{\Gamma^H_{GH^+} + \Gamma^H_{Cl^-}}{\Gamma^H_{GH^+} + \Gamma^H_{Cl^-} + \Gamma^H_{G^\pm}}. \tag{9.5}$$

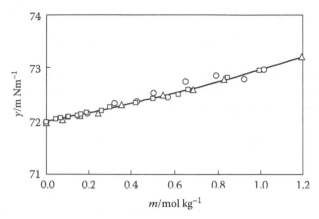

FIGURE 9.5   Surface tension–concentration curves of the binary mixtures of glycine and glycine hydrochloride.

In Figure 9.5, the surface tension of the binary mixtures of $G^{\pm}$ and GHCl is plotted against concentration under fixed X of 0.00, 0.53, and 1.00, respectively. The values of the GHCl solutions plotted in this figure are obtained by extrapolation of the γ values measured for HCl–GHCl mixtures to $X = 1$. This figure illustrates two interesting points. First, the contribution of $G^{\pm}$ that arises from dipole–water dipole interactions is consistent with the sum of the ion–water contribution of GH^{+} and Cl^{-} ions. Second, we have learned in the previous chapter that the additive property of the surface tension does not hold for the mixtures; especially, remarkable deviations are observed for anion–anion mixtures. The system under consideration can be considered as an anionic mixture with a common cation, $GH^{+}Cl^{-}$–$GH^{+}COO^{-}$, but no such deviation is observed.

9.2.2   Mixture of Glycine ($G^{\pm}$) and Sodium Glycinate

The surface tension data measured at fixed composition X are plotted against concentration m in Figure 9.6. The curves of mixtures are slightly concave upward. The slope increases as the fraction of NaG increases at low X, but the curve almost overlaps at high X. Here, m and X are defined as follows:

$$m = m_{G^{\pm}} + m_{NaG} \tag{9.6}$$

and

$$X = \frac{m_{NaG}}{m_{G^{\pm}} + m_{NaG}}. \tag{9.7}$$

Because of low solubility over X of 0.8, it is impossible to show γ at $X = 1$, but it is apparent that $d\gamma/dm$ of the NaG solution is larger than that of $G^{\pm}$. In contrast to the $G^{\pm}$–GHCl system, the surface tension of solutions obviously depends on X, and it deviates positively from the curve of the arithmetic mean of γ of single solute solutions. The system under consideration can be recognized as a cationic mixture with a

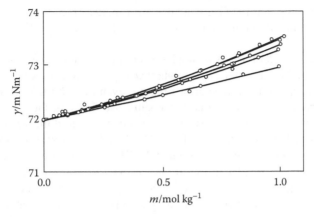

FIGURE 9.6 Surface tension–concentration curves of the binary mixtures of (1) glycine and (2) sodium glycinate at a fixed composition. From bottom to top, $X(2)$ are 0, 0.199, 0.396, 0.592, 0.799, respectively. A single line is used to connect the plots of the composition of 0.592 and 0.799.

common anion, GH^+COO^-–Na^+G^-. Equation 9.3 is applicable to the solutions under consideration if we define Γ^H and X^H as

$$\Gamma^H = \Gamma^H_{G^\pm} + \Gamma^H_{Na^+} + \Gamma^H_{G^-} \tag{9.8}$$

and

$$X^H = \frac{\Gamma^H_{Na^+} + \Gamma^H_{G^-}}{\Gamma^H_{G^\pm} + \Gamma^H_{Na^+} + \Gamma^H_{G^-}}. \tag{9.9}$$

Figure 9.7 shows the changes of $\Gamma^H_{G^\pm}$, $\Gamma^H_{G^-}$, and $\left(\Gamma^H_{G^\pm} + \Gamma^H_{G^-}\right)$ in X at the fixed surface tension of 72.5 mN m⁻¹, respectively. The $\Gamma^H_{G^\pm}$–X curve is obviously concave

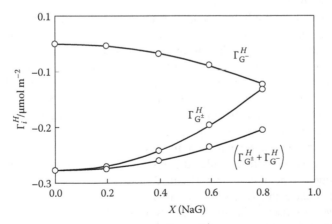

FIGURE 9.7 Variation of $\Gamma^H_{G^\pm}$, $\Gamma^H_{G^-}$, and $\left(\Gamma^H_{G^\pm} + \Gamma^H_{G^-}\right)$ with composition of $X(NaG)$ at the fixed surface tension of 72.5 mN m⁻¹.

upward, indicating that the $G^{\pm}$ molecules are repelled from the surface more than expected from the additivity rule. On the other hand, the $\Gamma_G^H - X$ curve is concave downward.

Now let us consider the distinction between dipolar molecules (zwitterionic) and 1:1 electrolytes. We are interested in whether dipolar molecules behave nearly the same as the simple salts in the surface region, even though their distance between the two charges is fixed. In the earlier results, we find that the solutions of zwitterionic glycine have $d\gamma/dm$ values nearly the same as that of the simple 1:1 valence type GHCl. Further, the solutions of $GH^+COO^-GH^+Cl^-$ mixtures also have nearly the same $d\gamma/dm$ values as those of $G^{\pm}$ and GH^+Cl^-. These results suggest that dipolar glycine is arranged parallel to the anion–cation pairs, and there are no particular dipole–charge interactions in the surface region. However, the following must be noted. In the anionic mixtures of simple salts with a common cation, the surface tension exhibits deviations from the arithmetic mean between the two that are always greater than those of cationic mixtures of salts with common anion. The inverse relationship exists between the simple salt mixtures and glycine mixtures.

In contrast to the $G^{\pm}$ and GH^+Cl^- mixtures, the deviation of the $d\gamma/dm - X$ curves of $G^{\pm}$ and Na^+G^- mixtures is observed (Figure 9.7), indicating that G^- ions are more attractive for dipolar water molecules than the zwitterionic glycine. Since the entropy data are not sufficient to allow a determination of the exact behavior of dipolar glycine in the mixture, the results indicate the possibility that the behavior of dipolar molecules in the surface region is not the same as those of simple salt.

9.3   ADSORBED FILM OF LEUCINE AND STRUCTURAL ISOMERS

The physical significance of the dipole–dipole interactions in the adsorbed film of zwitterionic leucine was briefly considered in Section 4.2.2. The surface tension versus concentration curves of the aqueous solution of leucine show two commonly recognized states of the adsorbed film, that is, gaseous and expanded states, and a clear transition point between them.

The size of the hydrophobic group of surfactants, as well as the type of hydrophilic group, generally plays a significant role in the adsorbed films. It is common knowledge that the change in the carbon number of a hydrophobic group strikingly changes the properties of the adsorbed film. The surface tension versus concentration curve of valine, 2-amino-3-methyl butanoic acid, which differs by only one in the number of carbon atoms with leucine, does not show the gaseous/expanded transition. So far, leucine is the only amino acid that shows transition of this type. The effect of interaction between dipole–dipole aligned parallel to the surface on the gaseous/expanded transition may be of interest.

In this section, we will consider the adsorbed film of four structural isomers: 2-amino-4-methyl pentanoic acid (Leu), 2-amino-3-methyl pentanoic acid (Ile), 2-amino-hexanoic acid (Nle), and 2-amino-3-dimethyl butanoic acid (Tle). And furthermore, we will discuss the mixed adsorbed film of Leu–Ile and Leu–Tle in order to provide the information about the effect of the isomeric structure on the adsorption at air/water surface.

Plots of surface tension versus concentration for the aqueous solutions of Leu, Ile, Nle, and Tle at 25°C are compared in Figure 4.11. These plots except for Tle can be divided into two straight lines. The lines beginning with zero concentration are graphs of the gaseous film, and the intersection between two straight lines is the gaseous/expanded transition point of the adsorbed film. The precise determination of this transition point is much more difficult than the determination of the value of $d\gamma/dm$ because the variation of surface tension is limited. However, an abrupt negative deviation from the linear plots for the gaseous adsorbed film indicates that transition occurs critically at the point, and the lateral interaction between dipoles is significant. The linear relationships between surface tension and concentration lead to an important conclusion that the partial molar thermodynamic quantity changes of adsorption remain unchanged. It is apparent that the solute–solute interactions are insignificant in the gaseous adsorbed film. Consequently, the magnitudes of the slope of surface tension versus concentration curves show differences in the magnitude of solute–solvent interactions between the surface region and bulk water. This is a property of all surface active substances, so-called surface activity, and may be evaluated immediately from the experimental data as

$$\frac{\Gamma^H}{m} = -\frac{1}{RT}\left(\frac{\partial \gamma}{\partial m}\right)_{T,p}. \tag{4.22}$$

The magnitudes of Γ^H/m can be considered a kind of partition coefficient characteristic of each surface active solute, because the observed values for dilute solutions remain constant at a given temperature and pressure. Figure 9.8 represents a graphical comparison of the Γ^H/m values evaluated for Leu, Ile, Nle, and Tle at 25°C. It is clear that Ile is less surface active than the other three isomers. Tle does not take gaseous/expanded transition in the concentration range from zero to the solubility limit, but Tle is rather more surface active than Ile in the whole concentration range. Nle whose hydrophobic group is a straight chain shows lower surface activity than Leu. The surface activity increases in the order Ile < Nle < Tle < Leu in the gaseous

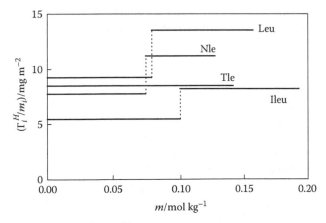

FIGURE 9.8 Comparison of $\left(\Gamma_i^H / m_i\right)$ of leucine, isoleucine, norleucine, and tert-leucine.

film and Ile < Nle < Leu in the expanded film. The difference in the surface activity of these isomers in the concentration range where they take the gaseous state may be attributable to the difference in the interactions between a hydrophobic group and water molecules, resulting from the difference in the size and structure of the hydrophobic group.

We next consider the properties of isomers in the adsorbed film. In the general discussion of the lateral interactions in a two-dimensional film, it has been shown advantageous to consider surface pressure π versus area per molecule A curves analogous to the pV-curves of three-dimensional systems. For the gaseous film of these isomers, the πA-plots lie on a single curve that represents the ideal gaseous film, since all γ–m relations observed are straight lines. For expanded films, the πA-plots also lie almost on a single curve, indicating that the magnitude of molecular interactions in the expanded films is practically insensitive to the structure of isomers or that the surface tension measurements are insensitive to changes in the chemical structure of leucine isomers in the film. However, it is important to note that although the differences in molecular interactions of leucine isomers in the film are not clearly presented by the πA-curves, there is distinct evidence of the presence of the effect of isomeric structure on the lateral interactions in the behavior of the transition point. The transition of Tle with a highly branched chain does not occur in the concentration region we examined, while that of Nle with a straight chain occurs at lower surface pressure and larger surface area than that of Leu whose chain terminal is branched. It is interesting that the transition of Ile with a straight ethyl group as chain terminal occurs at the point where the transition of Nle takes place. The earlier behaviors of the gaseous/expanded transition of leucine isomers are similar to that of boiling point observed for isomers in bulk phase. For example, it is well known that the boiling point of the highly branched isomers of alkanes or alcohols such as isopentane or t-amyl alcohol is lower than that of a straight-chain one.

The experimental data considered earlier show that the surface activity and the gaseous/expanded transition point are sensitive to the differences in molecular structure among isomers, although it is impossible to make a distinction between isomers on the πA-curves. In order to understand in detail the effect of the chemical structure of these isomers, we examined here the surface activity of Leu–Ile mixtures. The surface activity of the component i, Γ_i^H/m_i, in the mixtures of two isomers may be evaluated by

$$\frac{\Gamma_i^H}{m_i} = \frac{\Gamma_t^H X_i^H}{m_t X_i}. \tag{9.10}$$

In Figure 9.9, we have plotted Γ_t^H/m_t, $\Gamma_{\text{Leu}}^H/m_{\text{Leu}}$, and $\Gamma_{\text{Ile}}^H/m_{\text{Ile}}$ values of the gaseous film against X_{Ile}, respectively. As shown in Figure 9.8, the magnitude of the surface activity is primarily determined by the state of the adsorbed film irrespective of the change in concentration or in surface excess density. However, the values of X_i^H employed in Equation 9.10 are slightly dependent upon concentration, and we used the values of X_i^H at 70.5 mN m^{-1} for the gaseous state of the film. It is apparent that the plots of Γ_t^H/m_t versus X_{Ile} for the gaseous state of the films produce almost linear relationships connecting the surface activities of two pure isomers, though it seems slightly concave downward. The surface activity of Leu, $\Gamma_{\text{Leu}}^H/m_{\text{Leu}}$, is larger

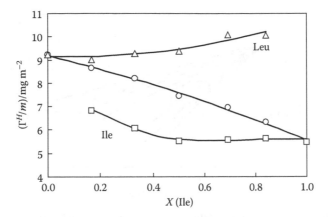

FIGURE 9.9 Surface activities in the mixed solution of leucine and isoleucine. Circles connected by a solid line show the surface activity of mixtures. Triangles and squares show that of leucine and isoleucine, respectively.

than that of Ile throughout the entire range of composition as is expected for ideal mixtures both in the bulk solution and in the surface region. The surface activity of Leu is almost constant especially for leucine-rich solutions, and the same is true for the surface activity of isoleucine.

9.4 ZWITTERIONIC LEUCINE ($L^\pm$) AND SODIUM LEUCINATE MIXTURES

Since the isoelectric point of leucine (2-amino-4-methylpentanoic acid) is 5.98, almost all of the leucine molecules in the aqueous solution have the form of zwitterion. By adding sodium hydroxide or hydrochloric acid into the aqueous leucine solution, the solution becomes a mixed solution of zwitterionic leucine ($L^\pm$) and sodium L-leucinate (NaL) or of $L^\pm$ and leucine hydrochloride (LHCl). We expect that surface tension measurements for these mixed solutions will provide us with information about the significance of the dipole–dipole interaction of leucine in the adsorbed film. In this section, we will consider the $L^\pm$–NaL mixtures to illustrate the distinct properties of zwitterions as the hydrophilic group.

Figure 9.10 shows γ–m curves for the $L^\pm$–NaL mixtures under fixed composition of NaL. The total concentration m and composition X of NaL are defined by

$$m = m_{L^\pm} + m_{Na^+} + m_{L^-} \tag{9.11}$$

and

$$X = \frac{m_{Na^+} + m_{L^-}}{m_{L^\pm} + m_{Na^+} + m_{L^-}}. \tag{9.12}$$

In this system, $L^\pm$ behaves as a weak acid with a dissociation constant of 2.51×10^{-10}. However, we expect no effect of H^+ and OH^- on the values of surface tension, since the concentration of both ions are negligible compared with the others. For $L^\pm$-rich mixtures, it is clearly seen that the γ–m relation is composed of two straight lines corresponding to

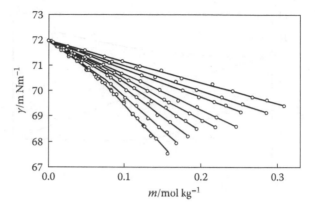

FIGURE 9.10 Surface tension of mixtures of leucine and sodium leucinate at constant composition and 25°C; from bottom to top compositions X are 0.00, 0.188, 0.366, 0.491, 0.596, 0.707, 0.819, 0.913, and 0.988, respectively.

the gaseous and expanded films as the same as that for solutions of $X = 0$. With increasing X, the breakpoint becomes unclear and disappears over $X = 0.7$. In Figure 9.11, the $-(d\gamma/dm)$ values of the gaseous and expanded films are plotted against X. The $d\gamma/dm$ value of the gaseous film varies slowly but uniformly along the entire composition range. On the other hand, the value of the expanded state film largely depends on the composition. The higher value indicates that there will be a more tightly packed film and stronger interactions in the expanded film than those in the gaseous film. The difference in hydrophilic groups between $L^\pm$ and NaL will explain these observations in Figures 9.10 and 9.11. Dipole–dipole interaction between zwitterions aligned parallel to the surface plays an important part in explaining why $d\gamma/dm$ of $L^\pm$ is larger than that of NaL and why $L^\pm$ shows critical gaseous/expanded transition in the film.

The variation of surface tension for the mixtures is given as a function of m and X:

$$d\gamma = -\frac{RT\Gamma^H}{m}dm + RT\Gamma^H\left[\frac{X - X^H}{X(1 - X)}\right]dX, \qquad (9.13)$$

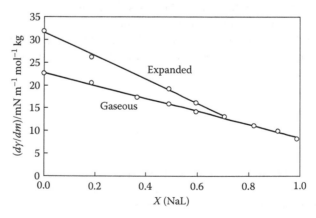

FIGURE 9.11 Variation of $-(d\gamma/dm)$ for the mixture of zwitterionic and anionic leucine.

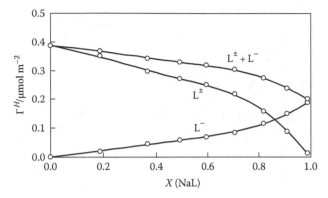

FIGURE 9.12 Adsorption of zwitterionic and anionic leucine from mixed solutions of leucine and sodium leucinate at the fixed surface tension of 71 mN m^{-1}.

where Γ^H and X^H are defined as

$$\Gamma^H = \Gamma_{L^\pm}^H + \Gamma_{Na^+}^H + \Gamma_{L^-}^H \tag{9.14}$$

and

$$X^H = \frac{\Gamma_{Na^+}^H + \Gamma_{L^-}^H}{\Gamma_{L^\pm}^H + \Gamma_{L^-}^H + \Gamma_{Na^+}^H}. \tag{9.15}$$

Let us consider the difference in the behavior of zwitterionic leucine and anionic leucine in the adsorbed film. These lateral interactions in the film bring about an alteration in the surface excess densities of the solutes. $\Gamma_{L^\pm}^H$, $\Gamma_{L^-}^H$, and $\Gamma_{L^\pm}^H + \Gamma_{L^-}^H$ calculated at fixed γ of 71 mN m^{-1} are plotted in Figure 9.12, respectively. This figure demonstrates clearly that $L^\pm$-rich films are formed throughout a wide range of composition. The curve of $\Gamma_{L^\pm}^H$ is concave to the X axis, and on the other hand the plot of $\Gamma_{L^-}^H$ is concave upward. Since the difference in the amount of the surface density evidently arises from the electrostatic effects of ion–ion, dipole–dipole, and ion-dipole interactions, the interaction between dipoles is more favorable than that between anions of L^-. As a result of these interactions, ionic leucines are depleted from the surface; in other words, there is a repulsive force between $L^\pm$ and L^- in the adsorbed film. These behaviors contrast Figure 9.7.

9.5   MIXED ADSORBED FILM OF LEUCINE AND SODIUM CHLORIDE

In the previous section, we have concluded that there are repulsive interactions between leucine and sodium leucinate in the adsorbed film. The 1:1 type sodium leucinate is repelled from the gaseous adsorbed film of leucine. This is opposite to the behavior of sodium glycinate. This behavior may be interpreted by the interaction between the dipole of the leucine and the ions of electrolytes. In this section, we consider the interaction between zwitterionic leucine lined up at the surface and ions of simple salts such as sodium chloride (Zhang and Matubayasi 2007).

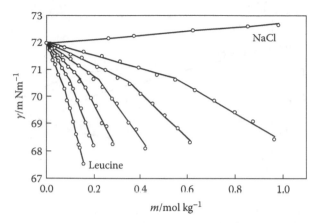

FIGURE 9.13 Surface tension–concentration curves of the leucine–sodium chloride mixtures. X are 1.00, 0.858, 0.773, 0.667, 0.508, 0.290, and 0.00 from top to bottom.

The characteristics of the leucine–sodium chloride mixtures are illustrated in Figure 9.13, in which the surface tension is plotted against total concentration at constant composition. Here, m and X are defined as follows:

$$m = m_{L^\pm} + m_{Na^+} + m_{Cl^-} \tag{9.16}$$

and

$$X = \frac{m_{Na^+} + m_{Cl^-}}{m_{L^\pm} + m_{Na^+} + m_{Cl^-}}. \tag{9.17}$$

Except for a pure NaCl solution, the curve shows clearly that the transition point is obtained as the intersection of two linear curves of the gaseous and expanded films. The transition points obtained in this manner have nearly the same surface tension values between 70.5 and 70.6 mN m^{-1}, although the total concentration m increases as the composition of NaCl increases. The graph of $d\gamma/dm$–X for the gaseous adsorbed film varies linearly and approaches zero at $X = 1$. It is probable that the ions of sodium chloride are immiscible in the adsorbed film and that there is no ion–dipole interaction in the adsorbed film.

Figure 9.14 shows the changes of $\Gamma_{L^\pm}^H$, $\left(\Gamma_{Na^+}^H + \Gamma_{Cl^-}^H\right)$, and Γ^H in X at the fixed surface tension of 71 mN m^{-1}, respectively, for comparison with Figure 9.12. Γ^H should be constant, and the small deviation from the horizontal line is due to the contribution of the activity coefficient. $\left(\Gamma_{Na^+}^H + \Gamma_{Cl^-}^H\right)$ decreases with increasing composition of sodium chloride, indicating that Na$^+$ and Cl$^-$ ions are repelled from the surface in a similar manner as a pure NaCl solution. The distinction between Figures 9.12 and 9.14 is apparent. In the former case, the 1:1 type sodium leucinate and leucine are freely miscible in the adsorbed film throughout the entire composition range. The surface excess density of the constituents versus composition diagram of the system is helpful for understanding the behavior of 1:1 surface active salts and zwitterions. In the latter case, however, it seems probable that leucine and sodium chloride are

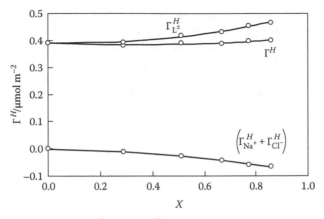

FIGURE 9.14 Variation of $\Gamma_{L^\pm}^H$, $\left(\Gamma_{Na^+}^H + \Gamma_{Cl^-}^H\right)$, and Γ^H with composition X.

not miscible in the film. In general, the surface pressure π versus area A curves are frequently used to illustrate the state of the adsorbed film, since accumulated data for insoluble films at the surface can be used as a useful reference. The graph of $\Gamma_{L^\pm}^H$ versus X shows that the area A of the adsorbed film of leucine decreases with increasing salt concentration when compared at a given surface pressure. The change of the film may be due to the variation of the chemical potential of leucine in the bulk phase along with the addition of simple salt.

In Figure 9.15, the Γ_{NaCl}^H/m_{NaCl} values at X of 0.508 are plotted against m_{NaCl}. Since the pure NaCl solution has a value of -0.313, the curve starts from a low value. However, the Γ_{NaCl}^H/m_{NaCl} value increases with increasing concentration, passes through an abrupt discontinuous drop at the gaseous/expanded transition point, and finally tends toward zero as concentration increases. The repulsion from the surface will be decreased along with the formation of the adsorbed film where dipoles are aligned parallel to the surface. The drop at the transition point may be due to the

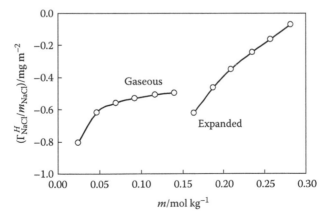

FIGURE 9.15 Variation of $\left(\Gamma_{NaCl}^H/m_{NaCl}\right)$ with total concentration of the mixture under the fixed composition of 0.508.

error in the determination of the transition point. With further decrease in X, the $\Gamma_{NaCl}^{H}/m_{NaCl}$ graph shifts in the direction of increasing m_{NaCl}, while the graph shifts in the direction of decreasing m_{NaCl} with further increase in X. It is to be noted that $\Gamma_{NaCl}^{H}/m_{NaCl}$ values of each mixed solution approach zero as concentration increases. However, it seems probable that sodium chloride is not mixed with leucine in the adsorbed film. The same behavior is observed for mixed solutions of isoleucine and sodium chloride (Zhang 2007).

REFERENCES

Belton, J. W. 1939. The capillary properties of α-amino acids. *Trans. Faraday Soc.* 35: 1293–1298.

Belton, J. W. and A. H. H. Twidle. 1940. The surface tensions of amino acid solutions as function of pH. *Trans. Faraday Soc.* 36: 1198–1208.

Bull, H. B. and K. Breese. 1974. Surface tension of amino acid solutions: A hydrophobicity scale of the amino acid residues. *Arch. Biochem. Biophys.* 161: 665–670.

Pappenheimer, J. R., M. P. Lepie, and J. J. Wyman. 1936. The surface tension of aqueous solutions of dipolar ions. *J. Am. Chem. Soc.* 58: 1851–1855.

Zhang, J. and N. Matubayasi. 2007 unpublished data.

Appendix

A.1 HISTORICAL BACKGROUND OF THE THERMODYNAMIC STUDIES ON THE SURFACE TENSION OF ELECTROLYTE SOLUTIONS

In the early twentieth century, there were many measurements that explained the general tendency of the effect of simple electrolytes on the air/water interface. Heydweiller (1910) reported the linear surface tension–concentration relations of moderately concentrated solutions of the order of mol kg^{-1} for more than 20 electrolytes. Langmuir (1917) has designed a surface region in which the surface layer should consist of a single layer of molecules of the solvent from which solute molecules are excluded. Langmuir wrote that "we see that if the surface tension increases linearly with the concentration, there is a deficiency of the solute in the surface layer which is proportional to the concentration." If the surface layer consists of pure water, the thickness τ of the solvent layer at the surface is calculated by

$$\tau = -\frac{\Gamma_i^G}{c_i} = \frac{1}{RT}\left(\frac{d\gamma}{dc_i}\right)$$

where c_i is a molarity. He compared some measurements for potassium chloride solutions and concluded that the layer of pure water adsorbed at the surface is about 0.4 nm thick. He tried to evaluate the thickness for other salts and obtained almost the same values. Harkins and McLaughlin (1925) and Harkins and Gilbert (1926) have determined the variation of this thickness with the change in the concentration of the added salts. Then, Schwenker (1931) has reported that salts of the same valence type gave almost identical increments of the surface tension. Based on these studies, three general ideas have been proposed for the aqueous salt solutions:

1. A linear surface tension–concentration relation is common for solutions of the order of mol kg^{-1}.
2. The increments of surface tension are primarily dependent upon the valence type assigned to the electrolytes.
3. There is a deficiency of ions in the surface region, and the topmost layer will be an ion-free layer.

Wagner (1924) and Onsager and Samaras (1934) have proposed a theoretical γ–concentration expression based on the model that the repulsion of the ions from the surface arises from the electrostatic image force. They have derived an expression from the Maxwell–Boltzmann equation for the adsorption of ions. The adsorption is expressed by

$$-\Gamma_i = \int_0^\infty \left[c - c(x) \right] dx \qquad (A.1)$$

Here, $c(x)$ is the concentration of ions at a distance x from the surface and is given by the Maxwell–Boltzmann equation:

$$c(x) = c \exp\left[-\frac{W(x)}{kT} \right] \qquad (A.2)$$

In order to obtain the expression for the work $W(x)$ required to bring an ion from the interior of a solution to a point at a depth x from the boundary, they employed the method of electrical images. Upon integration of Equation A.1, they have shown that the surface tension of the aqueous solution of 1:1-type electrolytes at 25°C is given by

$$\frac{\gamma}{\gamma^0} = 1 + \frac{1.012c}{\gamma^0} \log \frac{1.467}{c} \qquad (A.3)$$

The values computed by this equation and the values for KCl observed by Heydweiller (1910) are compared in Figure A.1. The concordance of theoretical and experimental results is excellent at dilute concentrations below 0.05 mol kg^{-1}. However, the theoretical curve is concave to the concentration axis, and departure from the experimental linear plot becomes significant as concentration increases. The linear variation of the surface tension seems strange, as many physical properties of the aqueous electrolyte solutions

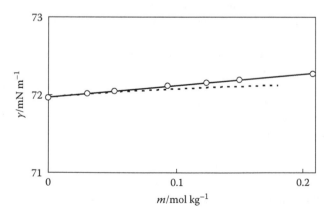

FIGURE A.1 Comparison of the theoretical surface tension–concentration relation for 1:1 electrolyte (dotted line) derived by Onsager and Samaras (1934) and experimental curve for KCl solution (circles) observed by Heydweiller (1910).

are plotted against the square root of concentration. A short review of the adsorption of simple salt solutions is given by Adam (1938).

The experimental data of surface tension are well cited in the textbook of colloid science by Jones and Ray (1937), because of their careful measurements and their anomalous γ–concentration curve. They tried to obtain an accurate surface tension value in a dilute concentration range of 0–0.01 mol L^{-1} and to compare it with the theoretical prediction. Their measurements agreed fairly well with the theoretical prediction except the anomaly known as the Jones–Ray effect. When the surface tension was measured as a function of increasing concentration in the very dilute solution, it first decreased with increasing concentration, passed through a small minimum, and then increased along with the theoretical curve. However, it was suggested that experimental conditions caused by a zeta potential of the quartz surface are responsible for such an anomaly in the γ–concentration curve (Langmuir 1938). An attempt to obtain the γ–concentration curves for dilute solutions below 0.01 mol L^{-1} was made by Long and Nutting (1942). They confirmed that no minimum in the γ–concentration curve is observed, and the results agree well with the Onsager–Samaras prediction. As a first approximation, image force explains well the general trends in the experimental observations that the γ–concentration curve of the electrolyte solutions primarily determined by the valence type and that salts of the same valence type are repelled equally from the surface. However, the slope of the γ–concentration curves for simple electrolytes measured by Heydweiller shows values specific for individual electrolytes when measurements are made up to high concentrations such as 1 mol kg^{-1}.

Johnasson and Eriksson (1974) have made careful surface tension measurements for the aqueous solution of four simple salts: NaCl, KCl, NaI, and KI. They have confirmed that the surface tension is a linear function of concentration and that specific anion effects are to a large extent decisive for the surface tension increments while the effect of sodium and potassium ions is nearly coinciding. Further, they provided a valuable insight into how the width of the surface region varies depending on the adsorption of ions. For a binary electrolyte solution–vapor system, they proposed the fundamental equation valid for a Verschaffelt–Guggenheim choice of surface phase:

$$-d\gamma = \left(\frac{S^{\sigma}}{A} - \Gamma_1 \frac{S}{c_1} \right) dT + \left(\Gamma_2 - \Gamma_1 \frac{c_2}{c_1} \right) d\mu_2 \tag{A.4}$$

This equation is the same as Equation 1.60 provided that c_1^{β}, c_2^{β}, and s^{β} are zero. Gibbs suggested that $\Gamma_1 / \left(c_1^{\alpha} - c_1^{\beta} \right)$ represents the distance τ between the surface of tension and the dividing surface, which makes $\Gamma_1 = 0$. If we define the new dividing surface so that $\Gamma_2 = \tau' c_2$, we have

$$-d\gamma = (\tau' - \tau)c_2 d\mu_2$$

at a given temperature. Then, the value of $(\tau' - \tau)$, which represents the distance between two dividing surfaces, can be evaluated from the experimental $d\gamma/dm$ data.

The results revealed that the thickness is independent of the concentration, and iodine ion can approach the surface more than chlorine ion. They have evaluated the γ–T slope of 1 mol kg^{-1} solutions and found that the magnitude of $-(d\gamma/dT)$ is smaller than that of pure water. Aveyard and Saleem (1976) have also measured the γ–concentration curves for some alkali metal halides at the air/water and oil/water interfaces. They also confirmed that the values of $d\gamma/dm$ of the aqueous solution of alkali metal chlorides are similar in magnitude and constant in the concentration range 0–1 mol kg^{-1} but differ significantly for bromides and iodides.

Besides the electrostatic repulsive interaction between the image force and ions, hydration of ions has been focused on the individual properties of distinctive ions because the relationship between the properties of ions and the surface tension is more directly obtained from the empirical properties of hydration. Hey et al. (1981) compared some 1:1 electrolytes, NaF, KF, NaNO$_3$, and KNO$_3$, and showed the correlation between surface tension increments and the hydration of salts. Weissenborn and Pugh (1995, 1996) have reported $d\gamma/dc$ values for many electrolytes of 1:1, 1:2, 1:3, and 2:2 types. They argued that there is a good correlation between $d\gamma/dc$ and the entropies of hydration for chloride solutions. Thereafter, many individuals proceeded to show how and why surface tension varies with the concentration of simple salts: Ghosh et al. (1988), Fedorova and Ulitin (2007), Pegram and Record (2006, 2007), Thckermann (2007), Ali et al. (2008, 2009), Marcus (2010), Sadeghi et al. (2010), Drzymala and Lyklema (2012), and Slavchov and Novev (2012).

A.2   SURFACE POTENTIAL MEASUREMENTS

Frumkin (1924) and Randels (1957, 1963) measured the surface potentials of aqueous electrolyte solutions and suggested that ions are repelled from the surface depending upon the individual properties of ions. Anions are repelled less strongly than cations; if anions are arranged in the order of the surface potentials of solutions of salts, the order represents the familiar lyotropic series. They pointed out that anions having the lowest hydration energies should approach the surface of water, but cations are relatively independent of the hydration energy.

Jarvis and Scheiman (1968) also measured the surface potentials of salt solutions and illustrated the structure of the surface more clearly as follows. Water molecules at the air/water interface have a preferential orientation and form an electrical double layer at the surface with the outermost portion of the double layer being negative and the innermost part being positive. Anions have preferentially accumulated to the positive side of the double layer. They also measured the surface tension by using a Du Nouy ring tensiometer with a precision of ±0.1 mN m^{-1} and compared the data with surface potential data. They observed poor correlation between the surface tension and the surface potentials of simple salt solutions. They pointed out that the surface tension values arising from the intermolecular force of attraction will be different with the electrical properties of the surface molecules. Randles (1977) has reviewed the surface potential data and Durand-Vidal et al. (2000) have briefly summarized the existing experimental methods and results of the properties of electrolytes at interface.

The consideration of the structure of a pure water surface based on the surface potential measurements has been confirmed by the theoretical and experimental results. Molecular dynamic simulations suggest that there is an asymmetric orientational distribution of water molecules, and this orientation of water brings the differences between the free energy profiles for ions of opposite charge (Wilson and Pohorille 1991). Recently, the surface of aqueous solutions of NaF, NaCl, NaBr, and NaI has been examined using vibrational sum frequency spectroscopy by Raymond and Richmond (2004). They observed that the hydrogen bonding of water in the surface region depends on the anion species and concluded that anions can approach the surface and interact with water in the surface region. In order to explain the distribution of ionic species in the surface region, theoretical approaches have been developed on the basis of the Poisson–Boltzmann theory (Onuki 2008). The distribution of ions is evaluated and Hofmeister series of electrolyte solutions are presented (dos Santos et al. 2010). The difference in the properties of ions between structure maker and structure breaker is well revealed (dos Santos and Levin 2012).

It is now more than 100 years since Heydweiller (1910) reported a systematic study on the surface tension of the aqueous electrolyte solutions. It is now certain that there are ions in the surface region and the electrical double layer is formed.

A.3 EXPERIMENTAL PROCEDURES OF SURFACE TENSION MEASUREMENTS FOR SIMPLE SALT SOLUTIONS

Generally, there are several alternative methods of obtaining surface tension values. These include techniques that are based on the analysis of the shape of a droplet, such as the sessile drop and the pendant drop method, and of the balance between gravitational forces acting on the liquid and surface tension forces that pull the liquid upward, such as the capillary rise, the drop volume, the Wilhelmy plate, and the maximum bubble pressure method. These methods provide precise surface tension values within up to ± 0.05 mN m^{-1} under ordinary conditions (Padday 1969).

Harkins and Brown (1919) have declared that the values for the surface tension of water reported by different observers using different methods vary by more than 6%. Prior to the establishment of the drop volume method by them, it was considered that the capillary rise method was the only one for the determination of the surface tension. However, they have pointed out three difficulties of the method: (1) difficulty in obtaining capillary tubes of uniform bore, (2) difficulty in reading with high accuracy, and (3) the fact that viscous liquids and solutions of certain compounds that corrode the glass surface give incorrect results. In order to fix these inconveniences, they carried out a standardization of the drop volume method for the determination of surface tension with high precision.

The drop volume method has disadvantages when aging of the adsorbed film is expected, but has the advantages that (1) the equipment is small and simple, (2) it is easy to measure the volume of a drop, and (3) it is easy to maintain an equilibrium state. Wilhelmy plate method is a convenient one for quick and easy measurement of the surface tension and used most widely, but it needs special care when precision and accuracy are needed. The use of the Du Nouy ring method is very limited because of the difficulty in estimating the adsorption at the surface. The sessile and pendant drop methods may be applied to a wide spectrum of liquid and solution surfaces. These methods enable the researcher to measure the absolute values of surface tension, but have difficulty in measuring a small variation of the surface tension. More information about the theoretical bases of these methods and outlines of the equipments is available in the book of Adamson (1982) and the review of Padday (1969).

The surface tension of the aqueous solutions of simple electrolytes considered in this book is measured by the drop volume method. A 200 µL disposable pipette (Drummond Scientific) was fused to a 1 mL glass syringe, and the cut end of the tip was polished plain perpendicular to the side so that no flaw could be seen. The radius of syringes was calibrated using water and mercury by weighing drops, and finally the radius of the tips was calibrated using pure water by assuming that the water has a value of 71.96 mN m^{-1} at 25°C. The measuring syringe was attached to a micrometer, and the moving length for a drop was recorded. The correction factor (F) was taken from the table provided by Lando and Oakley (1967).

Our measurements of the aqueous solution of the simple salts are made mostly in the concentration range 0–1 mol kg^{-1} so that the difference in the surface tension values between pure water and the solution could exceed more than 1.00 mN m^{-1}. To obtain three significant figures, we paid attention to the following: (1) the water was prepared by repeated distillation. We have avoided the use of plastic tubes, because decomposed plastics eluted by the steam could not be removed by distillation. (2) The glassware is kept in sulfuric acid–dichromate solution, washed with pure water just before use, and dried in a hand-made simple drying oven with a heat lamp inside instead of a commercially available one. (3) To attain water–vapor equilibrium, a small quantity of solution and a piece of filter paper were placed at the bottom of the measuring cell. The filter papers were dried before use to avoid volatile contaminations. (4) A drop of about 98% of its final volume is produced and allowed to stand for at lease 3 min before slowly turn off the drop. This procedure is repeated three to four times to confirm the equilibrium. In order to avoid a scattering of values, the temperature should be kept constant through this procedure. Hysteresis of the contact angle should be avoided. It is not discernible to the eye, but there may be a difference between advancing and receding contact angles.

For solutions, we conducted the measurements with the following three points in mind. First, a glass dropping tip is not suitable for solutions of anionic surface active substances. In Figure A.2, the surface tension of the aqueous solution of glucose is plotted against concentration. The aqueous solution of glucose is a mixture of cyclic isomers and a small amount of chain forms. Although the chain form does not take the anionic form, the scattering of the data is observed when a glass tip is used (closed circles), while the plots can be connected by a linear line when a platinum tip is used (open circles). It is to be noted that the γ of the aqueous solution of sucrose, which is a disaccharide composed of glucose and fructose, can be measured by using a glass dropping tip. However, we could not reuse glassware since a trace of the adsorbed sucrose could not be decomposed by a sulfuric acid–dichromate solution. Second, when pure water is put at the bottom of the cell to keep equilibrium of the vapor pressure, isopiestic distillation occurs. This situation is a case when surface

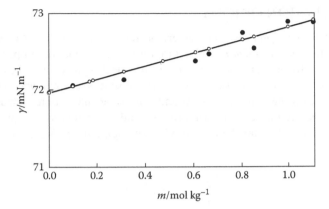

FIGURE A.2 Comparison of the surface tension–concentration relations measured by drop volume method using platinum (open circle) and glass (closed circle) tips.

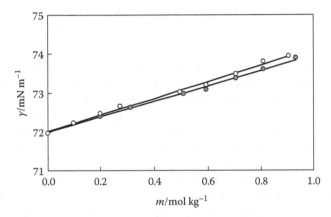

FIGURE A.3 Effect of isopiestic evaporation on the surface tension–concentration curve for calcium iodide solutions. Closed circles show the values of surface tension using drops in equilibrium with pure water, and open circles show the values in equilibrium with solutions of the same concentration.

tension is measured successively as a function of concentration using a single measuring cell. A typical example observed for the aqueous solution of calcium iodide is shown in Figure A.3. The open circles represent the case when the solution is put at the bottom of the cell and closed circles represent the case when pure water is used. It is observed experimentally that pure water moves to the pendant drop because of the difference in the vapor pressure between water and solution. More significant isopiestic distillation can be observed when a calcium chloride solution is used. Finally, it is to be noted that density data of the simple salt solutions reported in the literature are dependent on the observers in a similar manner to those of surface tension data. Thus, we have measured the densities of the solution with a few exceptions in order to change the probable error of a choice of the data to a systematic error accompanied by our own measurements.

A.4 HYDRATION OF IONS

When we consider $d\gamma/dm$ as a characteristic term of individual salt, we expect a positive relation between $d\gamma/dm$ and hydration of ions. Dynamic properties such as conductance, diffusion, and viscosity in a dilute electrolyte solution have been considered in view of ion–water interactions. The viscosity is employed commonly in the literature as a measure of ion–ion and ion–water interactions. Jones and Dole (1929) have tried to introduce a term proportional to the square root of the concentration into the equation for the relative viscosity, η/η_0, and proposed an equation of the form

$$\frac{\eta}{\eta_0} = 1 + A\sqrt{c} + Bc \tag{A.5}$$

which can be arranged to give

$$\frac{\eta/\eta_0 - 1}{\sqrt{c}} = A + B\sqrt{c} \tag{A.6}$$

A linear relation is obtained between $(\eta/\eta_0 - 1)/\sqrt{c}$ and $\sqrt{c}$ for a variety of electrolytes with some exceptions. Then the coefficients A and B have fixed values for a given salt solution that can be obtained graphically in a simple way. Debye and Hückel suggested that the interionic force opposing the motion of ions in an electric field is proportional to $\sqrt{c}$. Coefficient A represents the interionic contribution to the viscosity and has negative values. On the other hand, B is either negative or positive and is related to the ion–solvent interactions. Further, the coefficient B can be specified for a given ion under an assumption such that the B of potassium ion is equal to that of chlorine ion. Ions with positive B such as F^- ion are called structure makers and have larger hydration enthalpy than the ions called structure breakers such as I^- whose B values are negative.

Marcus (1985, 1987) has evaluated the standard molar enthalpies of hydration of an ion, ΔH_k, for the transfer of an ion from its gaseous state to the hydrated state. The standard states are the hypothetical ideal gas and ideal 1 M aqueous solution, respectively. He evaluated the values of ΔH_k for a lot of ions with considerable ingenuity using available thermodynamic data. As shown in Figure A.4, we can find close correlation between the ΔH_k and B coefficients of ions. For convenience of reference, the ΔH_k and B coefficients of the electrolytes in this book are shown in Table A.1.

A.5 HYDROLYSIS OF IONS

For purposes of interpretation of the experimental results, it sometimes is convenient to know the concentration of each constituent ion or molecule in equilibrium. In a chemistry textbook, many satisfactory explanations for the determination of ionization constants are available with its method of observation and calculation of the

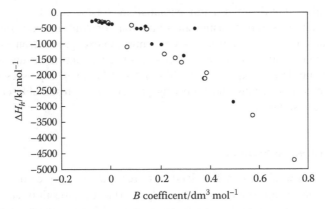

FIGURE A.4 Correlation between the hydration enthalpy and *B* coefficient of ions. Open circles: cations; filled circles: anions. (Data from Jenkins, H.D.B. and Marcus, Y., *Chem. Rev.*, 95, 2695, 1995; Marcus, Y., *J. Chem. Soc., Faraday Trans. 1*, 83, 339, 1987.)

TABLE A.1
Viscosity *B*-Coefficients and Enthalpies of Hydration of Ions

Ion	*B*-Coefficients (dm³ mol⁻¹)	Enthalpy (kJ mol⁻¹)	Ion	*B*-Coefficients (dm³ mol⁻¹)	Enthalpy (kJ mol⁻¹)
H^+	0.068	−1103	F^-	0.107	−510
Li^+	0.146	−531	Cl^-	−0.005	−367
Na^+	0.085	−416	Br^-	−0.033	−336
K^+	−0.009	−334	I^-	−0.073	−291
Rb^+	−0.033	−308	OH^-	0.122	−520
Cs^+	−0.047	−283	NO_3^-	−0.043	−312
NH_4^+	−0.008	−329	SO_4^{2-}	0.206	−1035
Mg^+	0.385	−1949	CrO_4^{2-}	0.165	−1012
Ca^+	0.284	−1602	MnO_4^-	−0.057	−250
Sr^+	0.261	−1470	CO_3^{2-}	0.294	−1397
Ba^+	0.216	−1332	ClO_4^-	−0.058	−246
Cu^+	0.376	−2123	ClO_3^-	−0.022	−299
Ni^+	0.379	−2119	BrO_3^-	0.009	−376
Al^+	0.744	−4715	IO_{3-}	0.14	−450
La^+	0.576	−3312	HPO_4^{2-}	0.382	
			$H_2PO_4^-$	0.34	−522
			PO_4^{3-}	0.495	−2879

Source: Data from Jenkins, H.D.B. and Marcus, Y., *Chem. Rev.*, 95, 2695, 1995; Marcus, Y., *J. Chem. Soc., Faraday Trans. 1*, 83, 339, 1987.

numerical values. On the other hand, explanations for the calculation of concentrations of the ions in solutions using an ionization constant are very limited. Since the ionization constant depends on the concentration except at extreme dilution, the calculations in the following are not strictly valid at concentrations up to 1 mol kg^{-1}. However, as the derived equation roughly and satisfactorily expresses the concentrations of ions in solutions, the following procedure is presumably convenient in consideration of solutions.

A.5.1 AMMONIUM HALIDES

When a salt of a weak acid is dissolved in water, changes in the concentrations of H$^+$ and OH$^-$ ions occur as a result of hydrolysis. If a sodium salt of acid HA is dissolved, for example, an acid anion A$^-$ reacts with a hydrogen ion to produce HA and OH$^-$ as

$$A^- + H_2O \rightleftharpoons HA + OH^-$$

with ionization constant

$$K_h = \frac{K_w}{K_A} = \frac{m_{HA} m_{OH^-}}{m_{A^-}} \frac{f_{HA} f_{OH^-}}{f_{A^-}}$$

Consequently, we have to take account of concentrations of ions and acid as a result of the hydrolysis if they are negligible or not. For a salt of a weak base, for example, on ammonium salt, the hydrolysis reaction is given as

$$NH_4^+ + H_2O \rightleftharpoons NH_3(aq) + H_3O^+$$

with ionization constant

$$K_h = \frac{K_w}{K_B} = \frac{m_{NH_3} m_{H^+}}{m_{NH_4^+}} \frac{f_{NH_3} f_{H^+}}{f_{NH_4^+}}$$

Let us consider the concentration of the chemical species in the aqueous solution of ammonium halide NH$_4$X. For convenience in calculation, we will use molality instead of the activity of species. In the aqueous solution of NH$_4$X, there are five unknown concentrations m_{NH_3}, $m_{NH_4^+}$, m_{H^+}, m_{OH^-}, and m_{X^-}. The evaluation of these concentrations requires the combination of five equations:

1. Electrical neutrality in an ionic solution,

$$m_{NH_4^+} + m_{H^+} = m_{OH^-} + m_{X^-}$$

2. Mass balance equations,

$$m = m_{NH_3} + m_{NH_4^+} = m_{X^-}$$

3. Two ionization constants,

$$K_B = \frac{m_{NH_4^+} m_{OH^-}}{m_{NH_3}} \quad \text{and} \quad K_W = m_{H^+} m_{OH^-}$$

By solving these equations, we find an equation for m_{H^+}:

$$m = \frac{\left(m_{H^+}^2 - K_W\right)\left(m_{H^+} K_B + K_W\right)}{m_{H^+} K_W}$$

If we try to solve this equation for m_{H^+}, the equation must be a complicated function. However, we can easily find out the values of m_{H^+} at a given concentration m by using the Solver function in Excel. Concentrations of other species can be obtained by the following equations:

$$m_{NH_4^+} = \frac{m m_{H^+} K_B}{m_{H^+} K_B + K_W}$$

and

$$m_{NH_3} = \frac{m K_W}{m_{H^+} K_B + K_W}$$

For an aqueous solution of NH$_4$Cl, m, m_{NH_3}, $m_{NH_4^+}$, m_{H^+}, and m_{OH^-} are shown in Table A.2, in which 1.77×10^{-5} is used as K_B. The m_{NH_3} and m_{H^+} in a 1.0 mol kg^{-1} NH$_4$X solution are less than 2.4×10^{-5} mol kg^{-1}. It is probable that m_{NH_3}, m_{H^+}, and m_{OH^-} can be neglected.

A.5.2 ALKALI METAL SULFATES

Sulfuric acid is a strong acid, and the bisulfate anion HSO$_4^-$ is quickly formed in an aqueous solution. The bisulfate ion is a weak acid whose dissociation constant is 0.0103 (Harned and Owen 1958, p. 568). When an alkali metal M$_2$SO$_4$ is dissolved in an aqueous solution, hydrolysis reaction proceeds spontaneously:

$$SO_4^{2-} + H_2O \rightleftharpoons HSO_4^- + OH^-$$

TABLE A.2

Concentrations of the Species Involved in the Aqueous Ammonium Halide Solution

m	$m_{NH_4^+}$	$10^6\, m_{NH_3}$	$10^6\, m_{H^+}$	$10^9\, m_{OH^-}$
0.20	0.20	11	11	0.95
0.40	0.40	15	15	0.67
0.60	0.60	19	19	0.55
0.80	0.80	21	21	0.47
1.00	1.00	24	24	0.42

In this solution, there are five ionic species $m_{SO_4^{2-}}$, $m_{HSO_4^-}$, m_{H^+}, m_{OH^-}, and m_{M^+} and the following five relations:

1. Electrical neutrality in an ionic solution,

$$m_{M^+} + m_{H^+} = m_{OH^-} + 2m_{SO_4^{2-}} + m_{HSO_4^-}$$

2. Mass balance equations,

$$m = m_{SO_4^{2-}} + m_{HSO_4^-} = 2m_{M^+}$$

3. Ionization constants,

$$K_A = \frac{m_{H^+}\, m_{SO_4^{2-}}}{m_{HSO_4^-}} \quad \text{and} \quad K_W = m_{H^+}\, m_{OH^-}$$

By solving these equations, the calculated values of $m_{HSO_4^-}$, $m_{SO_4^{2-}}$, m_{OH^-}, and m_{H^+} are shown in Table A.3. As expected from the equation of acid dissociation constant, the results indicate that $m_{SO_4^{2-}} \gg m_{HSO_4^-}$ in an alkaline solution.

A.5.3 AMMONIUM SULFATE

It is of our interest to find out the concentration of each species in aqueous $(NH_4)_2SO_4$ solutions. In this solution, there are six unknown concentrations $m_{NH_4^+}$, m_{NH_3}, $m_{SO_4^{2-}}$, $m_{HSO_4^-}$, m_{H^+}, and m_{OH^-}. For these unknowns we have six relations as

1. Electrical neutrality in an ionic solution,

$$m_{NH_4^+} + m_{H^+} = +2m_{SO_4^{2-}} + m_{HSO_4^-} + m_{OH^-}$$

TABLE A.3
Concentrations of the Species
Involved in the Aqueous Alkali Metal
Sulfate Solution

m	$m_{SO_4^{2-}}$	$10^6\, m_{HSO_4^-}$	$10^9\, m_{H^+}$	$10^6\, m_{OH^-}$
0.20	0.20	0.43	22	0.45
0.40	0.40	0.62	16	0.63
0.60	0.60	0.76	13	0.77
0.80	0.80	0.88	11	0.89
1.00	1.00	0.98	10	1.00

2. Two mass balance equations,

$$m = m_{SO_4^{2-}} + m_{HSO_4^-} \quad \text{and} \quad 2m = m_{NH_3} + m_{NH_4^+}$$

3. Three dissociation constants,

$$K_A = \frac{m_{H^+} m_{SO_4^{2-}}}{m_{HSO_4^-}}, \quad K_B = \frac{m_{NH_4^+} m_{OH^-}}{m_{NH_3}}, \quad \text{and} \quad K_W = m_{H^+} m_{OH^-}$$

The calculated values of $m_{NH_4^+}$, m_{NH_3}, $m_{SO_4^{2-}}$, $m_{HSO_4^-}$, m_{H^+}, and m_{OH^-} are shown in Table A.4. Since $K_A > K_B$, an acidic solution is formed.

A.5.4 GLYCINE

An ionization equilibrium of the zwitterionic glycine is described by two acid–base reactions:

$$NH_3^+CH_2COOH \leftrightarrow NH_3^+CH_2COO^- + H^+ \quad K_1 = 4.47 \times 10^{-3}$$

TABLE A.4
Concentrations of the Species Involved in the
Aqueous Ammonium Sulfate Solution

m	$m_{SO_4^{2-}}$	$10^3\, m_{HSO_4^-}$	$m_{NH_4^+}$	$10^3\, m_{NH_3}$	$10^6\, m_{H^+}$	$10^9\, m_{OH^-}$
0.20	0.20	0.048	0.40	0.053	4.3	2.3
0.40	0.40	0.098	0.80	0.10	4.4	2.3
0.60	0.60	0.15	1.20	0.15	4.5	2.3
0.80	0.80	0.20	1.60	0.20	4.5	2.3
1.00	1.00	0.25	2.00	0.25	4.5	2.3

and

$$NH_3^+CH_2COO^- \leftrightarrow NH_2CH_2COO^- + H^+ \quad K_2 = 1.66\times10^{-10}$$

The concentrations of the zwitterionic ($G^{\pm}$), cationic (G^+), and anionic (G^-) types are determined by three conditions:

1. Electrical neutrality in an ionic solution,

$$m_{G^+} + m_{H^+} = m_{G^-} + m_{OH^-}$$

2. Mass balance equation,

$$m = m_{G^+} + m_{G^-} + m_{G^{\pm}}$$

3. Three dissociation constants,

$$K_1 = \frac{m_{G^{\pm}}m_{H^+}}{m_{G^+}}, \quad K_2 = \frac{m_{G^-}m_{H^+}}{m_{G^{\pm}}}, \quad \text{and} \quad K_W = m_{H^+}m_{OH^-}$$

By solving these relations, m is given by the expression

$$m = \frac{(m_{OH^-} - m_{H^+})(m_{H^+}^2 + K_1m_{H^+} + K_1K_2)}{m_{H^+}^2 - K_1K_2}$$

The three types of glycine are

$$m_{G^-} = \frac{(m_{OH^-} - m_{H^+})K_1K_2}{m_{H^+}^2 - K_1K_2}$$

$$m_{G^+} = \frac{(m_{OH^-} - m_{H^+})m_{H^+}^2}{m_{H^+}^2 - K_1K_2}$$

and

$$m_{G^{\pm}} = \frac{(m_{OH^-} - m_{H^+})K_1m_{H^+}}{m_{H^+}^2 - K_1K_2}$$

TABLE A.5

Concentrations of the Species Involved in the Aqueous Glycine Solutions

m	$m_{G^\pm}$	$10^3\ m_{G^-}$	$10^3\ m_{G^+}$	$10^6\ m_{H^+}$	$10^9\ m_{OH^-}$
0.20	0.20	0.039	0.038	0.85	0.012
0.40	0.40	0.078	0.077	0.86	0.012
0.60	0.60	0.116	0.115	0.86	0.012
0.80	0.80	0.155	0.154	0.86	0.012
1.00	1.00	0.193	0.192	0.86	0.012

These values are shown in Table A.5 in which the value of $m_{G^\pm}$ is more significant than in others. The result indicates that the solution has a pH known as isoelectric point of glycine because of the electrical neutrality condition.

A.6   TABLES OF SURFACE TENSION OF ELECTROLYTES

As a unique title of this book, we have tabulated raw surface tension data in this section (Tables A.6 through A.40). We believe that a careful study of the surface tension will be helpful for anyone concerned with salt solutions in the wide field of application.

TABLE A.6

Surface Tension γ/mN m^{-1} of Lithium Chloride LiCl Solutions

m (mol kg^{-1})	Temperature (°C)								
	15.0	**17.5**	**20.0**	**22.5**	**25.0**	**27.5**	**30.0**	**32.5**	**35.0**
0.0000	73.58	73.18	72.79	72.38	71.98	71.58	71.18	70.78	70.38
0.0999	73.64	73.29	72.93	72.51	72.08	71.71	71.30	70.92	70.48
0.1047	73.65	73.24	72.88	72.51	72.09	71.69	71.28	70.87	70.46
0.2037	73.80	73.44	73.08	72.66	72.23	71.89	71.48	71.07	70.69
0.2940	73.93	73.54	73.15	72.79	72.36	71.90	71.58	71.19	70.81
0.4057	74.13	73.71	73.41	72.98	72.61	72.17	71.82	71.38	70.96
0.5031	74.22	73.78	73.46	73.08	72.68	72.29	71.85	71.51	71.04
0.6009	74.43	74.01	73.65	73.31	72.93	72.47	72.14	71.73	71.31
0.7010	74.46	74.12	73.72	73.34	72.95	72.54	72.17	71.83	71.41
0.8012	74.56	74.22	73.89	73.49	73.11	72.70	72.32	71.93	71.54
0.9031	74.75	74.39	74.02	73.68	73.31	72.87	72.54	72.13	71.73
0.9980	74.91	74.55	74.18	73.80	73.43	73.05	72.63	72.24	71.88

TABLE A.7
Surface Tension γ/mN m^{-1} of Sodium Chloride NaCl Solutions

	Temperature (°C)								
m (mol kg^{-1})	15.0	17.5	20.0	22.5	25.0	27.5	30.0	32.5	35.0
0.0000	73.58	73.18	72.79	72.38	71.98	71.58	71.18	70.78	70.38
0.0990	73.74	73.29	72.93	72.48	72.11	71.73	71.30	70.89	70.48
0.1982	73.86	73.41	73.08	72.65	72.25	71.81	71.43	71.05	70.67
0.2951	73.97	73.58	73.19	72.78	72.44	72.03	71.65	71.23	70.85
0.4012	74.14	73.74	73.37	72.97	72.56	72.18	71.77	71.38	70.99
0.4981	74.28	73.91	73.47	73.13	72.71	72.33	71.91	71.52	71.16
0.5937	74.46	74.08	73.70	73.31	72.89	72.52	72.15	71.72	71.29
0.7899	74.72	74.37	73.99	73.64	73.25	72.83	72.46	72.03	71.67
0.8389	74.83	74.45	74.04	73.68	73.32	72.93	72.56	72.16	71.79
1.0121	75.07	74.71	74.33	73.96	73.57	73.20	72.80	72.40	72.00

TABLE A.8
Surface Tension γ/mN m^{-1} of Potassium Chloride KCl Solutions

	Temperature (°C)								
m (mol kg^{-1})	15.0	17.5	20.0	22.5	25.0	27.5	30.0	32.5	35.0
0.0000	73.58	73.18	72.79	72.38	71.98	71.58	71.18	70.78	70.38
0.0994	73.69	73.34	72.98	72.58	72.18	71.72	71.38	70.87	70.55
0.2006	73.81	73.50	73.08	72.68	72.30	71.90	71.46	71.07	70.63
0.3006	73.94	73.61	73.21	72.86	72.46	72.02	71.69	71.24	70.80
0.3992	74.16	73.73	73.38	72.99	72.60	72.23	71.78	71.41	71.03
0.5012	74.28	73.90	73.54	73.13	72.74	72.34	71.97	71.54	71.19
0.5997	74.42	74.05	73.73	73.35	72.94	72.50	72.11	71.77	71.32
0.7007	74.56	74.26	73.90	73.54	73.17	72.77	72.31	71.91	71.45
0.7006	74.58	74.21	73.86	73.45	73.09	72.71	72.29	71.94	71.54
0.7988	74.76	74.41	74.05	73.63	73.26	72.89	72.48	72.13	71.74
0.9025	74.88	74.53	74.14	73.77	73.43	72.94	72.65	72.24	71.88
0.9990	75.02	74.72	74.27	73.96	73.51	73.13	72.80	72.42	72.04
0.1000	73.79	73.40	73.03	72.68	72.25	71.82	71.44	71.00	70.63

TABLE A.9
Surface Tension γ/mN m^{-1} of Cesium Chloride CsCl Solutions Between

	Temperature (°C)								
m (mol kg^{-1})	15.0	17.5	20.0	22.5	25.0	27.5	30.0	32.5	35.0
0.0000	73.58	73.18	72.79	72.38	71.98	71.58	71.18	70.78	70.38
0.1006	73.76	73.37	72.98	72.64	72.21	71.80	71.34	70.95	70.54
0.1989	73.88	73.52	73.15	72.75	72.31	71.96	71.54	71.12	70.73
0.2993	74.02	73.68	73.24	72.83	72.45	72.06	71.67	71.28	70.91
0.3968	74.13	73.75	73.41	73.02	72.63	72.21	71.85	71.39	71.02
0.5025	74.33	73.92	73.54	73.21	72.79	72.42	71.99	71.62	71.25
0.6066	74.47	74.09	73.73	73.37	72.98	72.61	72.20	71.80	71.42
0.7014	74.62	74.26	73.90	73.51	73.14	72.74	72.36	71.92	71.51
0.7960	74.72	74.37	74.00	73.67	73.26	72.86	72.48	72.13	71.71
0.8990	74.92	74.59	74.23	73.79	73.42	73.10	72.72	72.34	71.92
0.9895	75.09	74.75	74.39	74.01	73.63	73.25	72.87	72.48	72.09

TABLE A.10
Surface Tension γ/mN m^{-1} of Sodium Fluoride NaF Solutions

	Temperature (°C)								
m (mol kg^{-1})	15.0	17.5	20.0	22.5	25.0	27.5	30.0	32.5	35.0
0.0000	73.60	73.18	72.81	72.39	71.97	71.55	71.17	70.77	70.36
0.0942	73.74	73.35	72.94	72.49	72.08	71.64	71.22	70.85	70.45
0.0996	73.70	73.34	72.90	72.50	72.10	71.72	71.31	70.89	70.50
0.2211	73.92	73.54	73.19	72.78	72.34	71.90	71.47	71.10	70.73
0.3011	74.10	73.70	73.30	72.90	72.52	72.08	71.69	71.25	70.87
0.3280	74.18	73.70	73.34	72.93	72.54	72.14	71.69	71.27	70.86
0.3679	74.18	73.84	73.41	73.00	72.59	72.21	71.76	71.37	70.92
0.4860	74.39	74.01	73.62	73.23	72.81	72.41	71.98	71.57	71.19
0.5980	74.56	74.21	73.81	73.39	72.98	72.61	72.18	71.76	71.35
0.6859	74.74	74.35	73.99	73.56	73.11	72.73	72.32	—	71.52
0.7816	74.95	74.56	74.19	73.79	73.39	72.98	72.54	72.16	71.74
0.8735	75.06	74.69	74.35	73.92	73.52	73.11	72.70	72.33	71.91

TABLE A.11
Surface Tension γ/mN m^{-1} of Sodium Bromide NaBr Solutions

m (mol kg^{-1})	Temperature (°C)								
	15.0	17.5	20.0	22.5	25.0	27.5	30.0	32.5	35.0
0.0000	73.60	73.18	72.81	72.39	71.97	71.55	71.17	70.77	70.36
0.1006	73.67	73.28	72.89	72.49	72.09	71.68	71.27	70.83	70.41
0.1967	73.85	73.47	73.07	72.68	72.27	71.85	71.47	71.07	70.59
0.2990	73.95	73.54	73.16	72.77	72.40	71.97	71.55	71.14	70.73
0.4051	74.09	73.70	73.31	72.91	72.54	72.11	71.71	71.30	70.88
0.5016	74.20	73.79	73.38	73.04	72.67	72.27	71.84	71.43	71.00
0.6054	74.31	73.94	73.57	73.19	72.81	72.40	71.99	71.57	71.17
0.6975	74.44	74.07	73.67	73.29	72.88	72.49	72.04	71.65	71.22
0.6975	74.44	74.07	73.67	73.29	72.88	72.49	72.04	71.65	71.22
0.7829	74.47	74.10	73.76	73.35	72.95	72.56	72.20	71.78	71.39
0.8865	74.62	74.23	73.77	73.42	73.06	72.67	72.28	71.86	71.44
1.0049	74.85	74.45	74.09	73.71	73.32	72.93	72.53	72.11	71.68
1.0426	74.71	74.36	74.01	73.59	73.18	72.81	72.41	72.01	71.56

TABLE A.12
Surface Tension γ/mN m^{-1} of Sodium Iodide NaI Solutions

m (mol kg^{-1})	Temperature (°C)								
	15.0	17.5	20.0	22.5	25.0	27.5	30.0	32.5	35.0
0.0000	73.60	73.18	72.81	72.39	71.97	71.55	71.17	70.77	70.36
0.0855	73.65	73.23	72.85	72.50	72.08	71.69	71.30	70.90	70.42
0.1380	73.67	73.32	72.92	72.50	72.09	71.69	71.30	70.90	70.49
0.2497	73.85	73.46	73.08	72.68	72.29	71.89	71.48	71.08	70.64
0.2815	73.82	73.47	73.06	72.68	72.26	71.87	71.47	71.09	70.66
0.4846	74.00	73.66	73.26	72.88	72.50	72.08	71.69	71.27	70.87
0.5850	74.17	73.80	73.38	73.03	72.62	72.23	71.82	71.40	71.00
0.6643	74.20	73.85	73.48	73.12	72.68	72.35	71.93	71.50	71.11
0.7852	74.32	73.93	73.57	73.20	72.80	72.43	72.02	71.61	71.23
0.9654	74.48	74.12	73.73	73.36	72.98	72.58	72.20	71.85	71.43
1.0569	74.54	74.20	73.82	73.44	73.09	72.70	72.31	71.97	71.55

TABLE A.13
Surface Tension γ/mN m^{-1} of Lithium Nitrate LiNO$_3$ Solutions

	Temperature (°C)								
m (mol kg^{-1})	15.0	17.5	20.0	22.5	25.0	27.5	30.0	32.5	35.0
0.0000	73.58	73.18	72.79	72.38	71.98	71.58	71.18	70.78	70.38
0.0988	73.70	73.27	72.88	72.48	72.10	71.71	71.33	70.92	70.53
0.2001	73.79	73.38	72.99	72.61	72.22	71.82	71.44	71.06	70.65
0.2988	73.87	73.49	73.12	72.75	72.35	71.97	71.56	71.18	70.80
0.3994	74.00	73.62	73.25	72.84	72.46	72.04	71.67	71.25	70.87
0.5031	74.19	73.76	73.36	72.98	72.60	72.22	71.83	71.44	71.05
0.6003	74.24	73.85	73.47	73.10	72.69	72.29	71.88	71.50	71.12
0.7033	74.33	73.95	73.57	73.17	72.79	72.41	72.02	71.64	71.23
0.8015	74.41	74.01	73.68	73.28	72.87	72.50	72.12	71.76	71.38
0.9024	74.57	74.17	73.79	73.43	73.03	72.62	72.24	71.89	71.47
1.0022	74.63	74.24	73.86	73.48	73.11	72.72	72.34	71.96	71.58

TABLE A.14
Surface Tension γ/mN m^{-1} of Sodium Nitrate NaNO$_3$ Solutions Between

	Temperature (°C)								
m (mol kg^{-1})	15.0	17.5	20.0	22.5	25.0	27.5	30.0	32.5	35.0
0.0000	73.58	73.18	72.79	72.38	71.98	71.58	71.18	70.78	70.38
0.1022	73.71	73.24	72.87	72.44	72.07	71.68	71.31	70.87	70.51
0.2013	73.77	73.37	72.99	72.58	72.22	71.84	71.45	71.05	70.67
0.2997	73.98	73.58	73.22	72.79	72.40	71.99	71.62	71.21	70.82
0.3970	74.10	73.65	73.24	72.87	72.49	72.10	71.71	71.32	70.92
0.5013	74.16	73.75	73.35	72.98	72.59	72.23	71.79	71.45	71.03
0.5959	74.20	73.80	73.41	73.01	72.64	72.24	71.83	71.48	71.11
0.6997	74.24	73.92	73.54	73.19	72.77	72.44	71.99	71.68	71.22
0.8010	74.42	74.07	73.63	73.28	72.91	72.53	72.10	71.76	71.36
0.8999	74.54	74.13	73.75	73.36	73.00	72.67	72.26	71.90	71.51
1.0053	74.63	74.21	73.82	73.47	73.12	72.73	72.37	71.98	71.62

TABLE A.15

Surface Tension γ/mN m^{-1} of Potassium Nitrate KNO$_3$ Solutions

	Temperature (°C)								
m (mol kg^{-1})	15.0	17.5	20.0	22.5	25.0	27.5	30.0	32.5	35.0
0.0000	73.58	73.18	72.79	72.38	71.98	71.58	71.18	70.78	70.38
0.1007	73.70	73.27	72.87	72.49	72.08	71.70	71.31	70.91	70.52
0.2005	73.77	73.37	73.01	72.62	72.22	71.86	71.46	71.09	70.72
0.2984	73.89	73.48	73.12	72.71	72.33	71.95	71.59	71.19	70.83
0.4015	74.00	73.61	73.22	72.80	72.45	72.08	71.67	71.31	70.94
0.4981	74.07	73.68	73.25	72.92	72.54	72.16	71.80	71.41	71.06
0.6005	74.16	73.75	73.38	73.01	72.64	72.24	71.92	71.53	71.13
0.7010	74.24	73.84	73.51	73.10	72.76	72.38	72.03	71.63	71.23
0.7963	74.41	73.98	73.63	73.23	72.83	72.45	72.11	71.71	71.35
0.9019	74.49	74.14	73.72	73.34	73.00	72.58	72.24	71.82	71.47
0.9710	74.51	74.17	73.77	73.40	73.05	72.63	72.29	71.93	71.52

TABLE A.16

Surface Tension γ/mN m^{-1} of Ammonium Nitrate NH$_3$NO$_3$ Solutions

	Temperature (°C)								
m (mol kg^{-1})	15.0	17.5	20.0	22.5	25.0	27.5	30.0	32.5	35.0
0.0000	73.58	73.18	72.79	72.38	71.98	71.58	71.18	70.78	70.38
0.1001	73.66	73.26	72.85	72.47	72.02	71.67	71.28	70.89	70.49
0.2587	73.77	73.36	72.98	72.57	72.18	71.79	71.40	71.03	70.64
0.3888	73.90	73.52	73.14	72.76	72.33	71.94	71.54	71.18	70.78
0.5489	74.03	73.68	73.29	72.90	72.51	72.11	71.71	71.34	70.94
0.6794	74.11	73.76	73.37	72.98	72.61	72.21	71.81	71.44	71.07
0.8021	74.25	73.89	73.49	73.10	72.77	72.33	71.99	71.59	71.18
0.9013	74.36	73.96	73.60	73.23	72.83	72.43	72.06	71.65	71.27
0.9603	74.31	74.02	73.62	73.22	72.85	72.48	72.10	71.69	71.32
1.0426	74.41	73.98	73.65	73.31	72.94	72.54	72.20	71.79	71.41

TABLE A.17
Surface Tension γ/mN m^{-1} of Ammonium Chloride NH$_3$Cl Solutions

	Temperature (°C)								
m (mol kg^{-1})	15.0	17.5	20.0	22.5	25.0	27.5	30.0	32.5	35.0
0.0000	73.58	73.18	72.79	72.38	71.98	71.58	71.18	70.78	70.38
0.0501	73.61	73.24	72.86	72.45	72.03	71.67	71.24	70.85	70.46
0.1003	73.65	73.28	72.87	72.51	72.12	71.73	71.34	70.95	70.56
0.2007	73.80	73.44	73.01	72.62	72.22	71.87	71.47	71.07	70.70
0.3015	73.93	73.56	73.15	72.77	72.36	72.01	71.58	71.19	70.80
0.4110	74.04	73.69	73.28	72.90	72.52	72.13	71.74	71.31	70.95
0.4935	74.19	73.78	73.37	73.03	72.61	72.25	71.86	71.43	71.07
0.6048	74.31	73.94	73.55	73.18	72.73	72.37	71.98	71.58	71.19
0.7682	74.53	74.13	73.74	73.35	72.93	72.58	72.17	71.79	71.39
0.9192	74.67	74.28	73.90	73.52	73.09	72.77	72.34	71.94	71.58
0.9795	74.79	74.38	74.00	73.62	73.23	72.83	72.44	72.07	71.67

TABLE A.18
Surface Tension γ/mN m^{-1} of Ammonium Bromide NH$_3$Br Solutions

	Temperature (°C)								
m (mol kg^{-1})	15.0	17.5	20.0	22.5	25.0	27.5	30.0	32.5	35.0
0.0000	73.58	73.18	72.79	72.38	71.98	71.58	71.18	70.78	70.38
0.0955	73.58	73.28	72.87	72.42	72.13	71.64	71.25	70.88	70.45
0.1844	73.82	73.39	73.01	72.6	72.21	71.83	71.45	71.08	70.68
0.2776	73.93	73.55	73.14	72.79	72.34	71.98	71.58	71.19	70.82
0.4044	74.07	73.66	73.31	72.89	72.47	72.11	71.71	71.35	70.93
0.4959	74.17	73.79	73.38	73.02	72.6	72.24	71.84	71.48	71.07
0.6058	74.27	73.89	73.47	73.12	72.69	72.33	71.93	71.56	71.15
0.7056	74.36	73.93	73.57	73.17	72.84	72.47	72.06	71.72	71.27
0.7290	74.46	74.03	73.64	73.25	72.83	72.45	72.04	71.64	71.26
0.8022	74.49	74.1	73.71	73.34	72.9	72.53	72.16	71.77	71.38
0.8046	74.49	74.09	73.72	73.28	72.91	72.61	72.2	71.89	71.43
0.8882	74.59	74.22	73.85	73.48	73.1	72.69	72.27	71.85	71.54
0.8949	74.59	74.19	73.79	73.39	72.98	72.61	72.23	71.86	71.47
0.9648	74.64	74.26	73.9	73.49	73.19	72.74	72.35	71.93	71.55

TABLE A.19
Surface Tension γ/mN m^{-1} of Ammonium Iodide NH$_3$I Solutions

m (mol kg^{-1})	Temperature (°C)								
	15.0	17.5	20.0	22.5	25.0	27.5	30.0	32.5	35.0
0.0000	73.58	73.18	72.79	72.38	71.98	71.58	71.18	70.78	70.38
0.0984	73.63	73.22	72.85	72.43	72.02	71.63	71.24	70.88	70.45
0.2022	73.72	73.31	72.86	72.51	72.09	71.74	71.34	70.94	70.54
0.3929	73.86	73.44	73.05	72.66	72.22	71.89	71.49	71.09	70.68
0.4810	73.81	73.49	73.13	72.66	72.26	71.86	71.56	71.08	70.70
0.5925	73.92	73.56	73.16	72.76	72.32	71.98	71.61	71.19	70.78
0.6940	73.99	73.59	73.18	72.78	72.37	71.99	71.65	71.23	70.85
0.7912	74.08	73.75	73.34	72.93	72.52	72.17	71.79	71.37	70.99
0.9745	74.14	73.74	73.35	72.97	72.56	72.20	71.85	71.42	71.03

TABLE A.20
Surface Tension γ/mN m^{-1} of Calcium Nitrate Ca(NO$_3$)$_2$ Solutions

m (mol kg^{-1})	Temperature (°C)								
	15.0	17.5	20.0	22.5	25.0	27.5	30.0	32.5	35.0
0.0000	73.58	73.18	72.79	72.38	71.98	71.58	71.18	70.78	70.38
0.1001	73.73	73.34	72.97	72.57	72.19	71.78	71.40	71.01	70.59
0.1959	73.93	73.58	73.24	72.79	72.46	72.06	71.69	71.31	70.87
0.3033	74.17	73.82	73.46	73.05	72.72	72.32	71.86	71.51	71.13
0.4667	74.53	74.19	73.74	73.43	73.04	72.71	72.27	71.88	71.48
0.5084	74.59	74.21	73.89	73.50	73.14	72.69	72.36	71.91	71.58
0.5833	74.74	74.34	74.05	73.61	73.26	72.86	72.52	72.07	71.69
0.6845	74.97	74.67	74.26	73.82	73.54	73.06	72.76	72.29	72.00
0.7766	75.11	74.75	74.43	74.06	73.70	73.37	72.92	72.59	72.21
0.9000	75.43	75.08	74.68	74.32	73.95	73.60	73.14	72.76	72.42
0.9523	75.50	75.13	74.80	74.43	74.09	73.58	73.32	72.86	72.55

TABLE A.21
Surface Tension γ/mN m^{-1} of Calcium Chloride CaCl$_2$ Solutions

	Temperature (°C)								
m (mol kg^{-1})	15.0	17.5	20.0	22.5	25.0	27.5	30.0	32.5	35.0
0.0000	73.58	73.18	72.79	72.38	71.98	71.58	71.18	70.78	70.38
0.1170	73.91	73.54	73.19	72.74	72.38	71.99	71.63	71.19	70.75
0.1958	74.07	73.71	73.29	72.93	72.54	72.11	71.71	71.29	70.90
0.2786	74.29	73.91	73.50	73.04	72.80	72.36	71.96	71.57	71.14
0.3866	74.49	74.18	73.77	73.32	73.01	72.55	72.26	71.83	71.40
0.4718	74.75	74.46	74.03	73.73	73.29	72.92	72.51	72.20	71.74
0.5967	75.16	74.83	74.43	74.03	73.72	73.30	72.92	72.39	72.08
0.6846	75.40	75.07	74.63	74.28	73.87	73.48	73.14	72.68	72.36
0.7571	75.57	75.32	74.90	74.56	74.11	73.76	73.30	73.02	72.59
0.9083	75.96	75.61	75.30	74.95	74.56	74.14	73.82	73.35	72.91
0.9893	76.15	75.85	75.48	75.15	74.72	74.40	73.95	73.72	73.28

TABLE A.22
Surface Tension γ/mN m^{-1} of Calcium Bromide CaBr$_2$ Solutions

	Temperature (°C)								
m (mol kg^{-1})	15.0	17.5	20.0	22.5	25.0	27.5	30.0	32.5	35.0
0.0000	73.60	73.18	72.81	72.39	71.97	71.55	71.17	70.77	70.36
0.0515	73.66	73.30	72.92	72.50	72.13	71.76	71.36	70.93	70.56
0.0712	73.71	73.40	72.99	72.60	72.21	71.85	71.42	71.06	70.65
0.1427	73.91	73.56	73.14	72.78	72.35	71.99	71.59	71.18	70.77
0.1531	73.89	73.51	73.12	72.73	72.41	72.01	71.64	71.20	70.82
0.2146	74.06	73.74	73.35	72.95	72.55	72.14	71.77	71.33	70.91
0.2417	74.14	73.81	73.38	72.98	72.61	72.22	71.83	71.38	71.00
0.3308	74.35	73.90	73.59	73.20	72.84	72.44	72.10	71.64	71.22

TABLE A.23
Surface Tension γ/mN m^{-1} of Calcium Iodide CaI$_2$ Solutions

	Temperature (°C)								
m (mol kg^{-1})	15.0	17.5	20.0	22.5	25.0	27.5	30.0	32.5	35.0
0.0000	73.60	73.18	72.81	72.39	71.97	71.55	71.17	70.77	70.36
0.0975	73.75	73.40	72.98	72.59	72.23	71.83	71.46	70.99	70.61
0.2011	73.93	73.61	73.21	72.80	72.47	71.95	71.64	71.23	70.84
0.2765	74.11	73.74	73.37	72.96	72.65	72.12	71.88	71.50	71.04
0.3945	74.38	74.00	73.69	73.22	—	72.48	72.08	71.65	71.36
0.4966	74.54	74.22	73.82	73.42	73.02	72.62	72.25	71.88	71.51
0.5944	74.68	74.27	73.90	73.53	73.20	72.90	72.41	72.12	71.64
0.7050	74.92	74.62	74.16	73.86	73.48	73.10	72.71	72.25	71.98
0.9065	75.41	74.95	74.66	74.24	73.94	73.48	73.15	72.73	72.37

TABLE A.24
Surface Tension γ/mN m^{-1} of Magnesium Chloride MgCl$_2$ Solutions

	Temperature (°C)								
m (mol kg^{-1})	15.0	17.5	20.0	22.5	25.0	27.5	30.0	32.5	35.0
0.0000	73.58	73.18	72.79	72.38	71.98	71.58	71.18	70.78	70.38
0.0257	73.58	73.20	72.81	72.41	71.99	71.63	71.23	70.75	70.31
0.0537	73.70	73.29	72.90	72.50	72.15	71.72	71.31	70.88	70.51
0.0990	73.89	73.50	73.11	72.71	72.33	71.95	71.52	71.15	70.76
0.1961	74.13	73.74	73.34	72.96	72.56	72.17	71.81	71.42	70.97
0.2914	74.42	74.03	73.60	73.20	72.81	72.45	72.08	71.71	71.31
0.3849	74.65	74.27	73.91	73.48	73.14	72.75	72.43	71.98	71.58
0.4779	74.95	74.57	74.16	73.85	73.46	73.02	72.69	72.24	71.89
0.5644	75.22	74.83	74.37	74.03	73.64	73.24	72.86	72.51	72.18
0.6497	75.44	75.08	74.64	74.27	73.90	73.60	73.12	72.76	72.36
0.7343	75.73	75.37	74.97	74.58	74.25	73.77	73.39	73.06	72.72
0.8193	76.04	75.60	75.25	74.83	74.50	74.10	73.79	73.40	72.94

TABLE A.25
Surface Tension γ/mN m^{-1} of Sodium Phosphate Na$_3$PO$_4$ Solutions

m (mol kg^{-1})	Temperature (°C)								
	15.0	17.5	20.0	22.5	25.0	27.5	30.0	32.5	35.0
0.0000	73.58	73.18	72.79	72.38	71.98	71.58	71.18	70.78	70.38
0.0476	73.69	73.31	72.94	72.59	72.21	71.79	71.39	71.02	70.63
0.0981	73.82	73.47	73.07	72.74	72.29	71.96	71.52	71.22	70.81
0.1509	73.88	73.49	73.14	72.80	72.39	72.00	71.72	71.27	70.89
0.1754	74.04	73.62	73.33	72.91	72.53	72.12	71.68	71.31	70.98
0.1989	74.03	73.63	73.33	72.93	72.52	72.09	71.76	71.40	71.03
0.2516	74.19	73.79	73.41	73.11	72.63	72.36	72.01	71.62	71.24
0.2993	74.29	73.88	73.51	73.21	72.87	72.52	72.10	71.72	71.33
0.2996	74.23	73.85	73.56	73.23	72.76	72.42	72.06	71.75	71.35
0.3495	74.52	74.07	73.65	73.38	73.02	72.67	72.20	71.91	71.53
0.4014	74.58	74.16	73.81	73.42	73.07	72.67	72.31	72.02	71.59
0.4476	74.68	74.33	73.92	73.60	73.23	72.81	72.53	72.20	71.81

TABLE A.26
Surface Tension γ/mN m^{-1} of Sodium Hydrogen Phosphate Na$_2$HPO$_4$ Solutions

m (mol kg^{-1})	Temperature (°C)								
	15.0	17.5	20.0	22.5	25.0	27.5	30.0	32.5	35.0
0.0000	73.58	73.18	72.79	72.38	71.98	71.58	71.18	70.78	70.38
0.0471	73.63	73.23	72.82	72.40	72.05	71.68	71.30	70.84	70.42
0.1012	73.87	73.42	73.04	72.62	72.24	71.85	71.47	71.03	70.63
0.1434	73.87	73.52	73.10	72.71	72.35	71.91	71.53	71.14	70.83
0.2016	74.04	73.65	73.30	72.87	72.43	72.08	71.69	71.32	70.87
0.2287	74.05	73.66	73.33	72.93	72.53	72.14	71.78	71.40	70.99
0.2478	74.09	73.77	73.39	72.97	72.54	72.17	71.76	71.45	71.00
0.2965	74.27	73.81	73.48	73.04	72.67	72.19	71.88	71.47	71.08
0.3322	74.26	73.90	73.53	73.20	72.75	72.44	71.99	71.71	71.25
0.3969	74.50	74.10	73.62	73.31	72.86	72.56	72.15	71.79	71.40
0.4489	74.54	74.20	73.86	73.44	72.99	72.61	72.29	71.90	71.51

TABLE A.27
Surface Tension γ/mN m^{-1} of Sodium Dihydrogen Phosphate NaH$_2$PO$_4$ Solutions

m (mol kg^{-1})	15.0	17.5	20.0	22.5	25.0	27.5	30.0	32.5	35.0
				Temperature (°C)					
0.0000	73.58	73.18	72.79	72.38	71.98	71.58	71.18	70.78	70.38
0.0990	73.68	73.34	72.89	72.54	72.11	71.79	71.39	70.96	70.55
0.2019	73.92	73.54	73.12	72.73	72.34	71.97	71.54	71.20	70.76
0.2902	74.03	73.67	73.32	72.85	72.49	72.07	71.77	71.36	70.93
0.3992	74.27	73.81	73.52	73.05	72.77	72.27	72.02	71.44	71.14
0.5008	74.39	73.97	73.66	73.27	72.92	72.50	72.12	71.74	71.35
0.6006	74.62	74.18	73.81	73.44	73.09	72.71	72.33	71.94	71.55
0.6958	74.77	74.40	74.03	73.57	73.23	72.91	72.49	72.10	71.70
0.7918	74.95	74.50	74.20	73.85	73.43	73.07	72.64	72.34	71.85
0.8988	75.14	74.72	74.34	74.03	73.60	73.28	72.85	72.48	72.08
0.9915	75.27	74.93	74.51	74.15	73.78	73.44	73.08	72.67	72.27

TABLE A.28
Surface Tension γ/mN m^{-1} of Potassium Hydrogen Phosphate K$_2$HPO$_4$ Solutions

m (mol kg^{-1})	15.0	17.5	20.0	22.5	25.0	27.5	30.0	32.5	35.0
				Temperature (°C)					
0.0000	73.58	73.18	72.79	72.38	71.98	71.58	71.18	70.78	70.38
0.1016	73.74	73.44	73.02	72.61	72.18	71.83	71.47	71.04	70.63
0.1492	73.99	73.61	73.18	72.82	72.41	72.06	71.59	71.29	70.86
0.1978	74.07	73.67	73.28	72.84	72.47	72.10	71.74	71.29	70.96
0.2927	74.28	73.90	73.58	73.17	72.73	72.31	71.91	71.59	71.17
0.3351	74.37	74.01	73.64	73.28	72.87	72.47	72.10	71.68	71.32
0.4062	74.54	74.14	73.79	73.38	72.99	—	72.23	71.81	71.43
0.5038	74.85	74.47	74.08	73.68	73.28	72.91	72.51	72.16	71.77
0.5555	75.01	74.62	74.19	73.82	73.49	73.12	72.72	72.31	72.00
0.6020	75.12	74.71	74.39	73.94	73.58	73.13	72.83	72.46	72.05
0.6532	75.29	74.86	74.48	74.14	73.70	73.36	72.95	72.61	72.14
0.6993	75.34	74.98	74.59	74.29	73.81	73.51	73.05	72.75	72.29

TABLE A.29
Surface Tension γ/mN m^{-1} of Potassium Dihydrogen Phosphate KH$_2$PO$_4$ Solutions

m (mol kg^{-1})	Temperature (°C)								
	15.0	17.5	20.0	22.5	25.0	27.5	30.0	32.5	35.0
0.0000	73.58	73.18	72.79	72.38	71.98	71.58	71.18	70.78	70.38
0.0972	73.71	73.30	72.96	72.48	72.15	71.71	71.39	70.96	70.53
0.1959	73.91	73.50	73.17	72.71	72.32	71.95	71.55	71.16	70.76
0.3053	74.15	73.75	73.37	72.94	72.54	72.17	71.83	71.39	70.96
0.4068	74.29	73.90	73.50	73.15	72.71	72.32	71.97	71.58	71.17
0.5056	74.46	74.06	73.66	73.24	72.86	72.50	72.08	71.68	71.34
0.6026	74.63	74.27	73.83	73.45	73.02	72.65	72.26	71.90	71.50
0.6992	74.79	74.39	73.98	73.65	73.24	72.79	72.39	72.05	71.68
0.7965	74.93	74.56	74.11	73.75	73.36	72.99	72.59	72.26	71.78
0.8950	75.11	74.70	74.35	73.93	73.55	73.14	72.79	72.40	72.05
0.9626	75.27	74.80	74.40	74.07	73.73	73.24	72.96	72.47	72.17

TABLE A.30
Surface Tension γ/mN m^{-1} of Sodium Carbonate Na$_2$CO$_3$ Solutions

m (mol kg^{-1})	Temperature (°C)								
	15.0	17.5	20.0	22.5	25.0	27.5	30.0	32.5	35.0
0.0000	73.58	73.18	72.79	72.38	71.98	71.58	71.18	70.78	70.38
0.0362	73.62	73.23	72.87	72.49	72.10	71.68	71.28	70.86	70.47
0.1089	73.83	73.44	73.03	72.69	72.34	71.92	71.49	71.13	70.76
0.2192	74.13	73.75	73.36	72.98	72.55	72.20	71.76	71.40	70.90
0.2980	74.33	73.95	73.55	73.20	72.79	72.39	71.96	71.53	71.14
0.3379	74.45	74.06	73.67	73.27	72.87	72.47	72.12	71.69	71.30
0.3615	74.53	74.13	73.74	73.37	72.97	72.52	72.20	71.77	71.34
0.3950	74.55	74.24	73.86	73.45	73.11	72.67	72.29	71.85	71.47
0.4510	74.73	74.33	73.95	73.50	73.14	72.71	72.34	71.92	71.54
0.4639	74.74	74.35	73.94	73.54	73.17	72.77	72.41	72.00	71.56

TABLE A.31
Surface Tension γ/mN m^{-1} of Lithium Sulfate Li$_2$SO$_4$ Solutions

	Temperature (°C)								
m (mol kg^{-1})	15.0	17.5	20.0	22.5	25.0	27.5	30.0	32.5	35.0
0.0000	73.58	73.18	72.79	72.38	71.98	71.58	71.18	70.78	70.38
0.1011	73.82	73.46	73.07	72.66	72.24	71.87	71.46	71.07	70.70
0.2103	74.05	73.67	73.25	72.86	72.47	72.07	71.71	71.30	70.90
0.2566	74.12	73.80	73.34	72.98	72.52	72.22	71.75	71.45	70.90
0.4244	74.54	74.10	73.72	73.34	72.89	72.53	72.20	71.78	71.42
0.5114	74.73	74.36	73.98	73.58	73.18	72.76	72.40	72.00	71.61
0.6339	75.13	74.73	74.34	73.96	73.56	73.17	72.81	72.40	72.01
0.7592	75.34	74.96	74.54	74.16	73.70	73.39	73.07	72.69	72.26
0.9003	75.70	75.33	74.95	74.53	74.15	73.81	73.42	73.03	72.62
1.0038	75.99	75.64	75.22	74.87	74.48	74.09	73.70	73.30	72.95

TABLE A.32
Surface Tension γ/mN m^{-1} of Sodium Sulfate Na$_2$SO$_4$ Solutions

	Temperature (°C)								
m (mol kg^{-1})	15.0	17.5	20.0	22.5	25.0	27.5	30.0	32.5	35.0
0.0000	73.58	73.18	72.79	72.38	71.98	71.58	71.18	70.78	70.38
0.1345	73.79	73.42	73.00	72.62	72.20	71.81	71.42	71.02	70.65
0.1957	73.91	73.52	73.21	72.75	72.36	72.00	71.60	71.20	70.76
0.2152	74.01	73.63	73.24	72.82	72.42	71.99	71.59	71.19	70.75
0.3226	74.26	73.87	73.48	73.08	72.68	72.31	71.87	71.49	71.05
0.3490	74.25	73.82	73.46	73.09	72.76	72.25	71.84	71.43	71.05
0.4208	74.47	74.04	73.60	73.19	72.86	72.41	71.96	71.58	71.20
0.4584	74.61	74.19	73.82	73.38	72.99	72.67	72.14	71.86	71.36
0.5357	74.68	74.31	73.86	73.55	73.14	72.72	72.34	71.91	71.53
0.5580	74.72	74.34	74.00	73.62	73.24	72.79	72.44	72.01	71.59
0.6212	74.88	74.58	74.16	73.81	73.38	73.07	72.61	72.29	71.89
0.8027	75.12	74.66	74.31	73.99	73.63	73.23	72.87	72.40	71.96
0.9116	75.39	74.99	74.60	74.24	73.84	73.47	73.10	72.66	72.22
0.9975	75.65	75.25	74.95	74.56	74.13	73.79	73.49	73.11	72.70

TABLE A.33
Surface Tension γ/mN m^{-1} of Potassium Sulfate K$_2$SO$_4$ Solutions

	Temperature (°C)								
m (mol kg^{-1})	15.0	17.5	20.0	22.5	25.0	27.5	30.0	32.5	35.0
0.0000	73.58	73.18	72.79	72.38	71.98	71.58	71.18	70.78	70.38
0.0512	73.72	73.31	72.89	72.54	72.12	71.69	71.33	70.93	70.53
0.1010	73.84	73.42	73.00	72.61	72.19	71.79	71.43	71.03	70.66
0.1502	73.91	73.52	73.10	72.71	72.31	71.91	71.54	71.14	70.76
0.2027	74.07	73.64	73.25	72.85	72.46	72.05	71.72	71.31	70.94
0.2506	74.14	73.75	73.40	72.97	72.57	72.20	71.80	71.43	71.06
0.3432	74.36	73.95	73.56	73.08	72.71	72.30	72.06	71.62	71.29
0.3954	74.50	74.00	73.62	73.26	72.87	72.51	72.13	71.72	71.34
0.4407	74.63	74.15	73.78	73.37	72.97	72.58	72.29	71.82	71.55
0.4774	74.70	74.26	73.88	73.51	73.10	72.68	72.34	71.99	71.59
0.5978	74.94	74.53	74.10	73.70	73.36	72.89	72.58	72.22	71.86

TABLE A.34
Surface Tension γ/mN m^{-1} of Cesium Sulfate Cs$_2$SO$_4$ Solutions

	Temperature (°C)								
m (mol kg^{-1})	15.0	17.5	20.0	22.5	25.0	27.5	30.0	32.5	35.0
0.0000	73.58	73.18	72.79	72.38	71.98	71.58	71.18	70.78	70.38
0.1011	73.77	73.36	72.96	72.55	72.18	71.76	71.40	71.00	70.62
0.2012	74.04	73.62	73.27	72.86	72.47	72.15	71.65	71.29	71.03
0.2902	74.32	73.90	73.48	73.14	72.71	72.40	71.97	71.61	71.21
0.4036	74.54	74.14	73.79	73.39	73.03	72.67	72.26	71.90	71.53
0.5009	74.80	74.41	74.03	73.68	73.26	72.90	72.54	72.15	71.79
0.5996	74.98	74.60	74.29	73.89	73.51	73.17	72.79	72.44	72.06
0.7059	75.27	74.90	74.54	74.19	73.82	73.47	73.10	72.73	72.34
0.8005	75.48	75.17	74.77	74.46	74.06	73.75	73.39	73.03	72.71
0.8953	75.70	75.38	75.09	74.70	74.38	74.00	73.65	73.33	72.96
0.9972	76.01	75.64	75.31	74.99	74.63	74.29	73.96	73.67	73.30

TABLE A.35
Surface Tension γ/mN m^{-1} of Ammonium Sulfate (NH$_4$)$_2$SO$_4$ Solutions

	Temperature (°C)								
m (mol kg^{-1})	15.0	17.5	20.0	22.5	25.0	27.5	30.0	32.5	35.0
0.0000	73.58	73.18	72.79	72.38	71.98	71.58	71.18	70.78	70.38
0.0971	73.81	73.40	72.92	72.61	72.19	71.79	71.40	71.00	70.58
0.1986	73.93	73.62	73.24	72.81	72.44	71.99	71.59	71.19	70.82
0.3010	74.20	73.82	73.41	73.03	72.60	72.22	71.80	71.40	70.99
0.4950	74.49	74.17	73.75	73.37	72.97	72.56	72.18	71.74	71.32
0.5704	74.71	74.34	73.97	73.59	73.14	72.73	72.32	71.93	71.51
0.6963	74.95	74.58	74.21	73.81	73.40	73.00	72.57	72.17	71.80
0.8192	75.18	74.84	74.45	74.09	73.68	73.26	72.84	72.46	—
0.9117	75.33	74.99	74.57	74.23	73.83	73.43	72.98	72.58	72.21
1.0113	75.48	75.12	74.77	74.37	73.97	73.57	73.18	72.81	72.49

TABLE A.36
Surface Tension γ/mN m^{-1} of Magnesium Sulfate MgSO$_4$ Solutions

	Temperature (°C)								
m (mol kg^{-1})	15.0	17.5	20.0	22.5	25.0	27.5	30.0	32.5	35.0
0.0000	73.58	73.18	72.79	72.38	71.98	71.58	71.18	70.78	70.38
0.0486	73.62	73.29	72.87	72.49	72.12	71.74	71.37	70.99	70.57
0.1162	73.73	73.38	73.04	72.59	72.22	71.83	71.49	71.03	70.70
0.1208	73.74	73.39	73.05	72.63	72.26	71.82	71.41	71.07	70.70
0.1669	73.87	73.52	73.07	72.76	72.33	71.98	71.52	71.20	70.74
0.2162	73.95	73.54	73.21	72.77	72.43	72.05	71.58	71.26	70.80
0.2815	74.02	73.61	73.27	72.86	72.50	72.14	71.73	71.37	70.96
0.3501	74.19	73.75	73.34	73.04	72.61	72.28	71.86	71.50	71.07
0.3940	74.18	73.87	73.49	73.12	72.77	72.38	71.96	71.56	71.14
0.4578	74.31	73.94	73.64	73.17	72.82	72.44	72.10	71.67	71.31
0.4603	74.33	73.98	73.55	73.19	72.83	72.38	72.12	71.69	71.31
0.6031	74.49	74.13	73.74	73.35	73.01	72.62	72.24	71.86	71.46
0.6970	74.68	74.32	73.95	73.52	73.18	72.80	72.40	72.00	71.70
0.8054	74.83	74.49	74.07	73.76	73.36	73.02	72.64	72.28	71.79
0.9006	74.97	74.62	74.30	73.92	73.56	73.17	72.80	72.45	72.07
0.9988	75.15	74.77	74.47	74.09	73.73	73.32	72.96	72.61	72.22

TABLE A.37
Surface Tension γ/mN m^{-1} of Copper Sulfate CuSO$_4$ Solutions

	Temperature (°C)								
m (mol kg^{-1})	15.0	17.5	20.0	22.5	25.0	27.5	30.0	32.5	35.0
0.0000	73.58	73.18	72.79	72.38	71.98	71.58	71.18	70.78	70.38
0.1071	73.78	73.43	73.04	72.55	72.26	71.80	71.45	70.94	70.59
0.1957	73.97	73.56	73.18	72.78	72.33	71.99	71.55	71.17	70.78
0.2978	74.11	73.71	73.34	72.93	72.55	72.09	71.67	71.26	70.92
0.4472	74.43	74.01	73.70	73.28	72.78	72.27	71.97	71.52	71.21
0.5640	74.63	74.16	73.75	73.45	72.97	72.57	72.17	71.77	71.43
0.6555	74.78	74.38	73.99	73.54	73.12	72.73	72.32	71.91	71.59
0.7459	74.94	74.55	74.18	73.80	73.30	72.84	72.55	72.10	71.80
0.8134	74.98	74.63	74.26	73.83	73.39	72.98	72.62	72.24	71.78
0.9141	75.23	74.83	74.43	74.05	73.63	73.22	72.79	72.37	72.05

TABLE A.38
Surface Tension γ/mN m^{-1} of Nickel Sulfate NiSO$_4$ Solutions

	Temperature (°C)								
m (mol kg^{-1})	15.0	17.5	20.0	22.5	25.0	27.5	30.0	32.5	35.0
0.0000	73.58	73.18	72.79	72.38	71.98	71.58	71.18	70.78	70.38
0.1076	73.79	73.41	72.99	72.56	72.17	71.80	71.41	71.00	70.63
0.1986	73.94	73.53	73.15	72.72	72.32	71.95	71.51	71.14	70.74
0.1986	73.97	73.53	73.15	72.72	72.32	71.95	71.55	71.13	70.74
0.2831	74.08	73.72	73.28	72.89	72.49	72.11	71.68	71.27	70.92
0.4061	74.29	73.88	73.52	73.09	72.71	72.29	71.83	71.49	71.12
0.5041	74.49	74.08	73.62	73.28	72.89	72.43	72.00	71.65	71.29
0.6067	74.67	74.25	73.90	73.48	73.12	72.58	72.26	71.86	71.46
0.7080	74.84	74.45	74.02	73.63	73.23	72.75	72.39	72.02	71.62
0.8046	75.02	74.66	74.26	73.81	73.49	73.04	72.60	72.30	71.93
0.9016	75.21	74.84	74.39	74.02	73.57	73.16	72.75	72.37	72.07
0.9964	75.39	75.02	74.64	74.19	73.85	73.32	72.98	72.64	72.14

TABLE A.39
Surface Tension γ/mN m^{-1} of Some Sodium Salts at 25°C

NaMnO$_4$		NaClO$_3$		NaBrO$_3$		NaIO$_3$	
m (mol kg^{-1})	γ (mN m^{-1})	m (mol kg^{-1})	γ (mN m^{-1})	m (mol kg^{-1})	γ (mN m^{-1})	m (mol kg^{-1})	γ (mN m^{-1})
0.0000	71.96	0.0000	71.96	0.0000	71.96	0.0000	71.96
0.0498	72.04	0.0914	72.14	0.0999	72.07	0.0442	72.03
0.0928	72.09	0.2024	72.25	0.1992	72.30	0.1008	72.18
0.0996	72.09	0.3033	72.39	0.2976	72.47	0.1509	72.18
0.1492	72.18	0.3931	72.50	0.3949	72.60	0.1944	72.31
0.1808	72.27	0.4644	72.60	0.4977	72.76	0.2438	72.33
0.2465	72.38	0.5496	72.66	0.5587	72.80	0.2856	72.50
0.2465	72.38	0.6119	72.74	0.7028	73.08	0.3430	72.55
0.2984	72.41	0.6628	72.76	0.8132	73.16	0.3919	72.69
0.3012	72.48	0.7811	72.93	0.8632	73.25	0.4351	72.77
0.3538	72.49	0.8499	72.92	1.0173	73.36	0.4985	72.90
0.3979	72.61	0.9100	72.96	—	—	—	—
0.3979	72.54	0.9474	73.00	—	—	—	—
0.4458	72.68	0.9962	73.09	—	—	—	—
0.4575	72.70	—	—	—	—	—	—
0.4989	72.74	—	—	—	—	—	—

TABLE A.40
Surface Tension γ/mN m^{-1} of Some Potassium Salts at 25°C

KI		KBr		KMnO$_4$		NaClO$_4$	
m (mol kg^{-1})	γ (mN m^{-1})	m (mol kg^{-1})	γ (mN m^{-1})	m (mol kg^{-1})	γ (mN m^{-1})	m (mol kg^{-1})	γ (mN m^{-1})
0.0000	71.97	0.0000	71.97	0.0000	71.96	0.0000	71.96
0.1001	72.15	0.0846	72.06	0.0498	72.05	0.0924	72.00
0.1998	72.19	0.1986	72.25	0.1066	72.09	0.1975	72.01
0.2986	72.36	0.3032	72.37	0.1533	72.14	0.3015	72.08
0.3963	72.45	0.4110	72.51	0.1995	72.19	0.3925	72.08
0.4961	72.56	0.5157	72.68	0.2593	72.30	0.5190	72.10
0.6007	72.53	0.5224	72.72	0.3079	72.32	0.6937	72.12
0.6999	72.77	0.5762	72.80	0.3418	72.38	0.7910	72.10
0.8106	72.85	0.6355	72.84	—	—	0.8538	72.15
0.9012	73.00	0.7247	72.92	—	—	—	—
0.9873	73.04	0.7969	72.96	—	—	—	—
—	—	0.9294	73.18	—	—	—	—
—	—	1.0473	73.31	—	—	—	—

TABLE A.41
Graphically Smoothed [(dg/dm)/mN m^{-1} mol^{-1} kg] Values against Temperature

Salt	Temperature (°C)								
	15.0	17.5	20.0	22.5	25.0	27.5	30.0	32.5	35.0
LiCl	1.31	1.34	1.37	1.40	1.42	1.44	1.46	1.48	1.49
LiNO$_3$	1.06	1.09	1.11	1.14	1.16	1.18	1.19	1.21	1.22
Li$_2$SO$_4$	2.34	2.37	2.39	2.41	2.43	2.45	2.47	2.48	2.50
NaF	1.74	1.75	1.77	1.78	1.79	1.80	1.81	1.82	1.83
NaCl	1.46	1.49	1.51	1.53	1.55	1.57	1.59	1.60	1.61
NaBr	1.16	1.18	1.20	1.22	1.23	1.24	1.25	1.27	1.28
NaNO$_3$	1.03	1.07	1.10	1.13	1.16	1.18	1.20	1.22	1.24
NaI	0.93	0.96	0.99	1.02	1.04	1.07	1.09	1.11	1.12
NaH$_2$PO$_4$	1.68	1.73	1.76	1.80	1.83	1.85	1.88	1.90	1.92
Na$_2$HPO$_4$	2.17	2.23	2.29	2.34	2.38	2.42	2.46	2.50	2.53
Na$_3$PO$_4$	2.48	2.47	2.42	2.58	2.69	2.75	2.85	3.03	3.01
Na$_2$SO$_4$	2.02	2.04	2.06	2.07	2.09	2.10	2.11	2.12	2.13
Na$_2$CO$_3$[a]	2.53	2.58	2.58	2.62	2.73	2.63	2.70	2.65	2.62
KCl	1.46	1.49	1.52	1.55	1.57	1.59	1.61	1.63	1.65
KI	0.96	1.01	1.06	1.09	1.13	1.16	1.19	1.21	1.24
KNO$_3$	0.96	1.01	1.05	1.08	1.12	1.14	1.17	1.20	1.22
KH$_2$PO$_4$	1.70	1.72	1.74	1.76	1.78	1.79	1.80	1.82	1.83
K$_2$HPO$_4$	2.50	2.55	2.59	2.63	2.66	2.69	2.71	2.74	2.76
K$_2$SO$_4$	2.17	2.22	2.25	2.29	2.32	2.34	2.37	2.39	2.41
CsCl	1.50	1.54	1.57	1.60	1.63	1.65	1.68	1.70	1.72
Cs$_2$SO$_4$	2.40	2.49	2.57	2.64	2.71	2.76	2.81	2.86	2.91
NH$_4$Cl	1.22	1.23	1.24	1.25	1.26	1.27	1.27	1.28	1.29
NH$_4$Br	1.12	1.14	1.15	1.17	1.18	1.19	1.20	1.21	1.22
NH$_4$I	0.58	0.60	0.62	0.64	0.65	0.66	0.67	0.69	0.70
NH$_4$NO$_3$	0.81	0.84	0.88	0.90	0.93	0.95	0.97	0.99	1.01
(NH$_4$)$_2$SO$_4$	1.92	1.94	1.95	1.97	1.98	2.00	2.01	2.02	2.03
CaCl$_2$	2.65	2.69	2.73	2.76	2.79	2.82	2.85	2.87	2.89
CaBr$_2$	2.33	2.37	2.41	2.44	2.47	2.50	2.52	2.54	2.56
CaI$_2$	1.93	1.97	2.01	2.05	2.08	2.10	2.13	2.15	2.17
Ca(NO$_3$)$_2$	2.02	2.07	2.12	2.16	2.19	2.22	2.25	2.28	2.30
MgCl$_2$	2.88	2.92	2.96	2.99	3.02	3.05	3.08	3.10	3.12
MgSO$_4$	1.55	1.59	1.62	1.65	1.68	1.70	1.72	1.74	1.76
NiSO$_4$[a]	1.77	1.82	1.77	1.78	1.86	1.78	1.76	1.82	1.85
CuSO$_4$[a]	1.78	1.81	1.80	1.83	1.79	1.78	1.79	1.76	1.86

[a] The results are not smoothed.

We have been reported the $d\gamma/dm$ and γ_0 values as an empirical parameter of the linear regression lines. The average values of accumulated γ_0 are 73.58 mN m^{-1} (15.0°C), 73.18 (17.5), 72.79 (20.0), 72.38 (22.5), 71.98 (25.0), 71.58 (27.5), 71.18 (30.0), 70.78 (32.5), and 70.38 (35.0), respectively. These values are in good agreement with the mean values observed for pure water: 73.60 (15.0), 73.18 (17.5), 72.81 (20.0), 72.39 (22.5), 71.97 (25.0), 71.55 (27.5), 71.17 (30.0), 70.77 (32.5), and 70.36 (35.0). Using these values, we have evaluated the slope of $(\gamma - \gamma_0)$–concentration curves and compared with previously reported data. The most probable values of $d\gamma/dm$ are smoothed graphically against temperature, and values are shown in Table A.41.

A.7 SURFACE TENSION OF SUCROSE, GLUCOSE, AND FRUCTOSE

Tables A.42 through A.44 show the surface tension of the aqueous solution of sucrose, glucose, and fructose.

TABLE A.42
Surface Tension γ/mN m^{-1} of Sucrose

m (mol kg^{-1})	Temperature (°C)								
	15.0	17.5	20.0	22.5	25.0	27.5	30.0	32.5	35.0
0.0000	73.60	73.18	72.81	72.39	71.97	71.55	71.17	70.77	70.36
0.0979	73.60	73.23	72.89	72.45	72.12	71.70	71.27	70.81	70.41
0.1812	73.69	73.27	72.96	72.53	72.10	71.78	71.30	70.89	70.51
0.1992	73.74	73.29	72.97	72.46	72.13	71.68	71.26	70.85	70.52
0.2989	73.85	73.48	73.07	72.68	72.23	71.82	71.45	71.08	70.64
0.3974	73.93	73.58	73.16	72.76	72.30	71.96	71.47	71.14	70.76
0.5001	74.13	73.63	73.31	72.83	72.48	72.02	71.73	71.26	70.85

TABLE A.43
Surface Tension γ/mN m^{-1} of Glucose

m (mol kg^{-1})	Temperature (°C)								
	15.0	17.5	20.0	22.5	25.0	27.5	30.0	32.5	35.0
0.0000	73.60	73.18	72.81	72.39	71.97	71.55	71.17	70.77	70.36
0.1349	73.71	73.21	72.91	72.42	72.09	71.62	71.26	70.86	70.47
0.1679	73.73	73.30	72.93	72.57	72.13	71.78	71.35	70.94	70.53
0.2932	73.88	73.45	73.16	72.61	72.31	71.72	71.53	71.04	70.72
0.3296	74.01	73.57	73.06	72.64	72.28	71.78	71.50	71.03	70.70
0.4986	74.07	73.64	73.25	72.80	72.50	71.95	71.66	71.20	70.87
0.6601	74.24	73.87	73.44	73.06	72.66	72.20	71.82	71.46	71.05
0.8335	74.52	74.15	73.68	73.33	72.88	72.51	72.17	71.64	71.34
0.9121	74.55	74.10	73.74	73.36	73.02	72.60	72.25	71.78	71.43

TABLE A.44
Surface Tension γ/mN m^{-1}
of Fructose at 25°C

m (mol kg^{-1})	γ (mN m^{-1})
0.0000	71.96
0.0856	72.03
0.1746	72.23
0.2631	72.22
0.3591	72.38
0.4618	72.41
0.5461	72.53
0.6400	72.65

REFERENCES

Adam, N. K. 1938. *The Physics and Chemistry of Surfaces*, 2nd edn., Clarendou Press, Oxford, U.K.

Adamson, A. W. 1982. *Physical Chemistry of Surfaces*, 4th edn. Wiley, New York.

Ali, K., A. A. Shah, S. Bilal, and Azhar-ul-Haq. 2008. Thermodynamic parameters of surface formation of some aqueous salt solutions. *Colloids Surf. A Physicochem. Eng. Asp.* 330: 28–34.

Ali, K., A. A. Shah, S. Bilal, and Azhar-ul-Haq. 2009. *Colloid Surf. A Physicochem. Eng. Asp.* 337: 194–199.

Aveyard, R. and S. M. Saleem. 1976. Interfacial tensions at alkane—Aqueous electrolyte interfaces. *J. Chem. Soc., Faraday Trans. I* 72: 1609–1617.

Drzymala, J. and J. Lyklema. 2012. Surface tension of aqueous electrolyte solutions. Thermodynamics. *J. Phys. Chem. B.* 115: 12466–12472.

Durand–Vidal, S., J. P. Simonin, and P. Turq. 2000. *Electrolytes at Interfaces*. Kluwer Academic, Dordrecht, the Netherlands.

Fedorova, A. A. and M. V. Ulitin. 2007. Surface tension and adsorption of electrolytes at the aqueous solution–gas interface. *Russ. J. Phys. Chem. A* 81: 1124–1127.

Frumkin, A. Z. 1924. Phasengrenzkräfte und adsorption an der trennungsfläche luft Lösung anorganisheher electrolyte. *Z. Phys. Chem.* 109: 34–48.

Ghosh, L., K. P. Das, and D. K. Chattorj. 1988. Thermodynamics of adsorption of inorganic electrolytes at air/water and oil/water interfaces. *J. Colloid Interface Sci.* 121: 278–288.

Harkins, W. D. and F. E. Brown. 1919. The determination of surface tension (free surface energy), and the weight of falling drops: The surface tension of water and benzene by the capillary height method. *J. Am. Chem. Soc.* 41: 499–524.

Harkins, W. D. and E. C. Gilbert. 1926. The structure of films of water on salt solutions. II. The surface tension of calcium chloride solutions at 25°. *J. Am. Chem. Soc.* 48: 604–607.

Harkins, W. D. and H. M. McLaughlin. 1925. The structure of films of water on salt solutions I. Surface tension and adsorption for aqueous solutions of sodium chloride. *J. Am. Chem. Soc.* 47: 2083–2089.

Harned, H. S. and B. B. Owen. 1958. *The Physical Chemistry of Electrolytic Solutions*, 3rd edn., American Chemical Society Monograph series. Reinhold Publishing Corporation, New York.

Hey, M. J., D. W. Shield, J. M. Speight, and M. Will. 1981. Surface tensions of aqueous solutions of some 1:1 electrolytes. *J. Chem. Soc., Faraday Trans. I* 77: 123–128.

Heydweiller, A. 1910. Über physikalische eigenshaften von lösungen in ihrem zusammen-
    hang. II Overflächenspannung und elektrisches leitvermögen wässeriger salzlösungen.
    Ann. Physik. [4] 33: 145–185.
Jarvis, N. L. and M. A. Scheiman. 1968. Surface potentials of aqueous electrolyte solutions.
    J. Phys. Chem. 72: 74–78.
Jenkins, H. D. B. and Y. Marcus. 1995. Viscosity B-coefficients of ions in solution. *Chem. Rev.*
    95: 2695–2724.
Johnasson, J. and C. K. Eriksson. 1974. γ and $d\gamma/dT$ measurements on aqueous solutions of
    1, 1 – electrolytes. *J. Colloid Interface Sci.* 49: 469–480.
Jones, G. and M. Dole. 1929. The viscosity of aqueous solutions of strong electrolytes with
    special reference to barium chloride. *J. Am. Chem. Soc.* 51: 2950–2964.
Jones, G. and W. A. Ray. 1937. The surface tension of solutions of electrolytes as a function of
    the concentration. I. A differential method for measuring relative surface tension. *J. Am.
    Chem. Soc.* 59: 187.
Lando, J. L. and H. T. Oakley. 1967. Tabulated correction factors for the drop-weight volume
    determination of surface and interfacial tensions. *J. Colloid Interface Sci.* 25: 526–530.
Langmuir, I. 1917. Fundamental properties of solid and liquid. *J. Am. Chem. Soc.* 39:
    1848–1906.
Langmuir, I. 1938. Repulsive forces between charged surfaces in water, and the cause of the
    Jones-ray effect. *Science* 88: 430–432.
Long, F. A. and G. C. Nutting. 1942. The relative surface tension of potassium chloride solu-
    tions by a differential bubble pressure method. *J. Am. Chem. Soc.* 64: 2476–2482.
Marcus, Y. 1985. *Ion Solvation.* John Wiley & Sons, New York.
Marcus, Y. 1987. The thermodynamics of solvation of ions. Part 2—The enthalpy of hydration
    at 298.15 K. *J. Chem. Soc., Faraday Trans. 1.* 83: 339–349.
Marcus, Y. 2010. Surface tension of aqueous electrolytes and ions. *J. Chem. Eng. Data* 55:
    3641–3644.
Onsager, L. and N. N. T. Samaras. 1934. the surface tension of Debye-Hückel electrolytes.
    J. Chem. Phys. 2: 528–534.
Onuki, A. 2008. Surface tension of electrolytes: Hydrophilic and hydrophobic ions near an
    interface. *J. Chem. Phys.* 128: 224704-1–224704-9.
Padday, J. F. 1969. Surface tension II. In *Surface and Colloid Science*, ed. E. Matijevic,
    pp. 101–149. Wiley, New York.
Pegram, L. M. and M. T. Record Jr. 2006. Partitioning of atmospherically relevant ions between
    bulk water and the water vapor interface. *Proc. Natl Acad. Sci. USA* 103: 14278–14281.
Pegram, L. M. and M. T. Record Jr. 2007. Hofmeister salt effects on surface tension arise
    from partitioning of anions and cations between bulk water and the air-water interface.
    J. Phys. Chem. B. 111: 5411–5417.
Randles, J. E. B. 1957. Ionic hydration and the surface potential of aqueous electrolytes.
    Discuss. Faraday Soc. 24: 194–199.
Randles, J. E. B. 1963. The Interface between aqueous electrolyte solutions and the gas phase.
    In *Advances in Electrochemistry and Electrochemical Engineering*, eds. P. Delahay and
    C. W. Tobias, Vol. 3, pp. 1–30. Wiley Interscience, New York.
Randles, J. E. B. 1977. Structure at the free surface of water and aqueous electrolyte solutions.
    Phys. Chem. Liq. 7: 107–179.
Raymond, E. A. and G. L. Richmond. 2004. Probing the molecular structure and bonding of
    the surface of aqueous salt solutions. *J. Phys. Chem. B.* 108: 5051–5059.
Sadeghi, M., V. Taghikhani, and V. Ghotbi. 2010. Measurement and correlation of surface ten-
    sion for single aqueous electrolyte solutions. *Int. J. Thermophys.* 31: 852–859.
dos Santos, A. P. and Y. Levin. 2012. Ions at the water–oil interface: Interfacial tension of
    electrolyte solutions. *Langmuir* 28: 1304–1308.

Schwenker, G. 1931. Über eine wesentliche verfeinerung der oberflächenspannungsmessung nach der bügelmethode und über die overflächenspannung verdünnter salzlösungen. *Ann. Physik.* [5]11: 525–557.

Slavchov, R. I. and J. K. Novev. 2012. Surface tension of concentrated electrolyte solutions. *J. Colloid Interface Sci.* 387: 234–243.

Thckermann, R. 2007. Surface tension of aqueous solutions of water-soluble organic and inorganic compounds. *Atmos. Environ.* 41: 6265–6275.

Wagner, C. 1924. The surface tension of dilute solutions of electrolytes. *Physik. Z.* 25: 474–477.

Weissenborn, P. K. and R. J. Pugh. 1995. Surface tension and bubble coalescence phenomena of aqueous solutions of electrolytes. *Langmuir* 11: 1422–1426.

Weissenborn, P. K. and R. J. Pugh. 1996. Surface tension of aqueous solutions of electrolytes: Relationship with ion hydration, oxygen solubility, and bubble coalescence. *J. Colloid Interface Sci.* 184: 550–563.

Wilson, M. A. and A. Pohorille. 1991. Interaction of monovalent ions with the water liquid-vapor interface: A molecular dynamics study. *J. Chem. Phys.* 95: 6005–6013.

Index

Milton Keynes UK
Ingram Content Group UK Ltd.
UKHW040100071024
449327UK00019B/688